The Value of Social Media for Predicting Stock Returns

Michael Nofer

The Value of Social Media for Predicting Stock Returns

Preconditions, Instruments and Performance Analysis

With a Foreword by Prof. Dr. Oliver Hinz

 Springer Vieweg

Michael Nofer
Darmstadt, Germany

Dissertation, TU Darmstadt, Germany, 2014

ISBN 978-3-658-09507-9   ISBN 978-3-658-09508-6 (eBook)
DOI 10.1007/978-3-658-09508-6

Library of Congress Control Number: 2015935424

Springer Vieweg
© Springer Fachmedien Wiesbaden 2015

Printed on acid-free paper

Springer Vieweg is a brand of Springer Fachmedien Wiesbaden
Springer Fachmedien Wiesbaden is part of Springer Science+Business Media
(www.springer.com)

Foreword

Firms like Facebook or Pinterest that have access to a large number of users impact our daily life in many aspects. Facebook for example has over one billion users registered for their online service allowing their user base to manage social contacts, to post and share content and to communicate their taste by clicking on "Like" buttons that are nowadays available on many websites. Many agree that their market capitalization cannot be justified by their tangible assets like machines or inventories, but bases mainly on their access to unique consumer data that makes these firms special. Users of these networks leave digital footprints, reveal their preferences or make social recommendations that seem valuable for e-business.

On the one hand a lot of these data is freely accessible and on the other hand these data can be used to forecast future developments and can thereby potentially be monetized. This potential makes this field so attractive these days. Big Data is assumed to be the new oil and we are currently in the middle of a gold rush.

Researchers and practitioners alike believe that Internet data are also valuable for beating the stock market. Michael Nofer's dissertation tries to assess the value of social media for predicting stock returns and examines the preconditions, instruments and finally assesses the potential performance gains. It is one of the first dissertations that examines this phenomenon with such a great care, with such a huge data base and with a different set of methods.

A common theme of this book is the thoughtful approach in all essays in identifying the important and timely research questions and the depth at which the authors examines the issue at hand. This is not an easy undertaking and I laud the nice empirical work that has been carried out.

I highly recommend this book to both, practitioners and researchers who are interested in predicting the development of the stock market. The book has the potential to be one of the milestones in this domain and in my opinion the readers can highly benefit from Michael Nofer's work. I wish the author all the best with this publication and I believe that the book will be a huge success!

Technische Universität Darmstadt   Prof. Dr. Oliver Hinz

Acknowledgements

This dissertation was accepted by TU Darmstadt in November 2014. I would like to take this opportunity to thank those people who supported me during my time at the Chair of Information Systems | Electronic Markets.

I would particularly like to thank my supervisor, Prof. Dr. Oliver Hinz, for giving me the chance to conduct research under outstanding conditions in such an interesting field of study. When he founded the Chair in 2011, I was fortunate to be among the first PhD students. Despite his numerous responsibilities, Prof. Hinz fully dedicates his time to his students whenever needed. Every day I enjoyed working in such a pleasant atmosphere, gaining new insights again and again. It was also a great pleasure to see the team grow over the years. I additionally thank Prof. Dr. Alexander Benlian (Chair of Information Systems and Electronic Services, TU Darmstadt) for co-supervising my dissertation.

One article of this dissertation was written alongside Prof. Dr. Jan Muntermann (University of Göttingen) and Dr. Heiko Roßnagel (Fraunhofer-Gesellschaft). I am very grateful for the opportunity to work in conjunction with such notable researchers.

I would also like to thank my colleagues at TU Darmstadt. We have supported each other not only with technical knowledge, but with our interpersonal support. The advice I received during the scientific colloquiums was especially valuable to me. I particularly thank Markus Franz for comforting me after the defeats suffered by my favorite football club Karlsruher SC. I am also grateful to have met talented students who supported me in terms of data collection and programming tasks.

Finally, I want to thank my friends and especially my wonderful family for accompanying me on this journey. This dissertation would not have been possible without the encouragements of those people who unconditionally supported me over the years.

Darmstadt   Michael Nofer

Table of Contents

List of Figures

List of Tables

List of Abbreviations

AIC	Akaike information criterion
AIM	Affect infusion model
API	Application programming interface
ASTS	Aktuelle Stimmungsskala
BIC	Bayesian information criterion
EAN	European article number
ETF	Exchange-traded fund
FWB	Frankfurter Wertpapierbörse
ISIN	International Securities Identification Number
POMS	Profile of mood states
SAD	Seasonal affective disorder
SMI	Social Mood Index
UGC	User generated content
VIF	Variance inflation factor
WoC	Wisdom of Crowds
WSMI	Weighted Social Mood Index

1 Introduction

1.1 Synopsys

Social Media applications have gained increasing importance in recent years, thanks largely to broadband connections allowing people faster, easier access to the Internet. According to an eMarketer report, 1.97 billion people worldwide are using Social Media applications in 2014, and this number is expected to grow to 2.55 billion by the year 2017 (eMarketer 2013). The most prominent examples of Social Media websites include social networks (e.g., Facebook, Google+), blogs and microblogs (e.g., Twitter), content communities (e.g., Flickr, YouTube), and virtual worlds (e.g., Second Life, World of Warcraft). This dissertation builds upon the work of Kaplan and Haenlein (2010), who define Social Media as a "group of Internet-based applications that build on the ideological and techno-logical foundations of Web 2.0, and that allow the creation and exchange of User Generated Content."

Social Media contains a vast amount of user-generated content, which is primarily produced by private individuals rather than professionals pursuing commercial interests (Tirunillai and Tellis 2012). Furthermore, much of the content is publicly available and can be retrieved as well as analyzed with the help of data mining techniques (e.g., Bifet and Frank 2010). Academics and industry professionals routinely employ the phrase "big data" to refer to this exponential increase in information present on the Internet.

Due to its nature (i.e., vast, publicly available, and directly created by end users), Social Media provides an interesting data source for companies and researchers. In particular, individuals post valuable information about themselves to Social Media that can be used to forecast future events and business develop-ments. For instance, flu epidemics can be forecasted using Google search queries, as people search for flu-related terms at the beginning of an outbreak (Ginsberg et al. 2009). With respect to consumer behavior, researchers have successfully forecasted music sales using MySpace (Dhar and Chang 2009), movie sales using the Hollywood Stock Exchange (Spann and Skiera 2003), and book sales using Amazon (Chevalier and Mayzlin 2006).

Similarly, this dissertation examines whether user-generated content on So-cial Media platforms can be used to predict stock returns. This investigation is a response to a long-standing debate on whether share prices can be forecasted or not. According to the efficient market hypothesis, it is impossible to predict share prices since every piece of information is immediately factored in (Fama 1970). In the last few decades, however, researchers have repeatedly exposed market anomalies that contradict the efficiency of financial markets. For instance, Jaffe and Westerfield (1985) found that stock returns are higher on Monday compared to other days of the week. Inefficient markets are therefore a precondition for the prediction of share prices.

Overall, the dissertation comprises four published articles. The first article refers to market anomalies on Social Media platforms and deals with the question of whether electronic markets are efficient. Research on market efficiency in the area of Finance is transformed to the Internet. Understanding how Internet users process information is necessary for judging the efficiency of the aforementioned markets.

The second and third articles examine the predictive value of user-generated content with regard to stock returns. Data was collected directly from Social Media applications, which offer interesting possibilities for researchers and practitioners in the area of share price forecasting (e.g., Bollen et al. 2010). Previous studies, for example, used consumer reviews (e.g., Tirunillai and Tellis 2012) or discussions on stock message boards (Antweiler and Frank 2004) to forecast stock market developments. The second article specifically refers to the "Wisdom of Crowds" phenomenon and uses data from a stock prediction community, while the third article deals with whether mood states collected from Twitter users have predictive value for stock returns. Mood analysis aims to determine people's feelings and emotions, and among Internet users, such information can be extracted from social networks and microblogs such as Facebook or Twitter. This data can, in turn, serve as proxy for investors' risk appetite and thus their willingness to invest in stocks (Bollen et al. 2010; Karabulut 2011).

The fourth article investigates the influence of privacy and security incidents on consumer or investor behavior. Privacy and security have been identified as important preconditions for Internet users' willingness to share content on Social Media platforms (Fogel and Nehmad 2009). Internet users' trust in Social Media platforms can be shaken by data theft or hacker attacks, which occurred increasingly over the last years (Kelly 2013; Silveira 2012). However, researchers and practitioners need ongoing access to content from Social Media platforms in order to use this information for the prediction of share returns. It is therefore necessary to study people's behavior in case of privacy and security incidents, which possibly threaten people's willingness to communicate and thus the flow of information.

While the first three articles use data from Social Media applications, the fourth article describes the outcomes of a laboratory experiment that was conducted on a university campus. The remainder of this introduction is devoted to first elucidating the theoretical background and research contexts, and then summarizing each article.

The four articles are:

1) Nofer, Michael / Hinz, Oliver (2012). Market Anomalies on Two-Sided Auction Platforms. European Conference on Information Systems (ECIS), Barcelona, Spain.

2) Nofer, Michael / Hinz, Oliver (2014). Are Crowds on the Internet Wiser than Experts? – The Case of a Stock Prediction Community. Journal of Business Economics, 84 (3), 303-338.

3) Nofer, Michael / Hinz, Oliver (2014). Using Twitter to Predict the Stock Market: Where is the Mood Effect? Business & Information Systems Engineering, forthcoming.

4) Nofer, Michael / Hinz, Oliver / Muntermann, Jan / Roßnagel, Heiko (2014). The Economic Impact of Privacy Violations and Security Breaches – A Laboratory Experiment. Business & Information Systems Engineering, forthcoming.

1.2 Research Contexts

1.2.1 Market Efficiency

According to the efficient market hypothesis, financial markets are efficient in such a way that every piece of information is immediately reflected in market prices (Fama 1970). Theoretically, then, investors should be unable to achieve superior returns compared to the mean market averages in the long run. However, several anomalies have been empirically observed in financial markets. Prominent examples of these price distortions include calendar anomalies, such as the "weekend effect" (e.g., Fields 1931; Jaffe and Westerfield 1985), and technical anomalies, such as the "momentum effect" (Jegadeesh and Titman 1993). These anomalies contradict the assumption of market efficiency.

For share price forecasting, it is important to know whether information can be used as a predictor. If markets are efficient, then information should be immediately factored into current share prices, which entails that investors cannot use said information to achieve superior returns. The presence of market anomalies in financial markets extends investors' predictive powers, but it remains unclear whether such anomalies also exist on Internet platforms. This dissertation thus aims to close this gap in the literature by investigating transactions between buyers and sellers on an online auction platform.

1.2.2   Wisdom of Crowds

The Wisdom of Crowds (WoC) has gained increasing attention from researchers and practitioners since the term was introduced by popular science author James Surowiecki (2004). This idea basically contends that a large crowd can make better decisions than a few individuals. One prominent example in the offline world is the "Ask the Audience" lifeline of "Who Wants to Be a Millionaire", which refers to the correct answer in over 90 percent of all cases (Surowiecki 2004). The literature also points toward instances when a large crowd on the Internet can be used to solve complex tasks: Studies show, for example, that Wikipedia's entries are of high quality and possess an accuracy comparable to the expert-written entries in print encyclopedias (Giles 2005; Rajagopalan et al. 2011). It is hardly surprising, then, that companies are "crowdsourcing" tasks to the Internet, allowing dedicated individuals to assist with tasks like problem solving (Howe 2008).

In the area of finance, there exist so-called stock prediction communities on the Internet. The Wisdom of Crowds is apparent on these platforms, as many individuals from diverse backgrounds continually meet to discuss stocks. This dissertation utilizes the share recommendations of one community and compares them with those of professional analysts and the average market. The dissertation therefore extends the findings of earlier researchers, who also used data from financial discussion boards such as Raging Bull or Yahoo! Finance (e.g., Antweiler and Frank 2004; Tumarkin and Whitelaw 2001). These earlier studies already demonstrated that online discussions about stock returns possess predictive value. However, before the existence of sophisticated Social Media platforms, authors had to analyze the messages with the help of text mining techniques. In contrast, stock prediction communities now offer researchers to collect much more data, including stock picks based on buy and sell ratings.

Besides the performance analysis, another major goal is to study the preconditions for wise crowds on the Internet. In the offline world, four conditions for wise crowds have been identified: knowledge, motivation, diversity, and independence (Simmons et al. 2011). This dissertation will clarify how changing degrees of diversity and independence influence the performance of the crowd in an Internet setting. Diversity means that crowd members are diverse with respect to various characteristics such as knowledge, culture, age, gender, or education. Previous research has also highlighted the importance of crowd members' independent decisions. In the case of the Internet crowd, every participant can make a decision without being influenced by other group members.

1.2.3   Mood Analysis

The relationship between mood states and stock returns is well established in the literature (e.g., Kamstra et al. 2003). According to the misattribution bias, people's risk appetite depends on their mood states (Johnson and Tversky 1983). Thus, if investors are in a good mood, share prices tend to increase since investors are more willing to take risks. On the other hand, the risk appetite is lower if people are in bad mood, which leads to decreasing share prices.

A number of studies in finance have uncovered market anomalies that are driven by people's mood states. For instance, the weather effect refers to the phenomenon that share returns are higher on sunny days than on cloudy days (Hirshleifer and Shumway 2003; Saunders 1993). Kamstra et al. (2003) investigated the role of depression, finding that share returns are lower during autumn and winter when many individuals suffer from seasonal affective disorder. Sport events, sleeping habits, and air pollution can also influence investors' mood states and in turn affect the stock market (Edmans et al. 2007; Kamstra et al. 2000; Levy and Yagil 2011).

Today, Social Media applications allow researchers to measure mood states in real-time. Initial research indicates that mood states extracted from Twitter or Facebook can predict stock returns. This dissertation will extend previous findings in the literature by considering the community structure, which might play an important role when measuring mood states. Experimental research has shown that mood states can spread among individuals (Bono and Ilies 2006; Sy et al. 2005). Relatedly, a number of recent studies have observed emotional contagion occurring via text-based communication on the Internet (Guillory et al. 2011; Hancock et al. 2008; Kramer 2012). Using public data from Twitter, this dissertation will test whether emotional contagion on the Internet might affect people's offline behavior, particularly their investment behavior.

1.2.4   Privacy and Security

One fundamental condition for the existence of Social Media applications is the willingness of Internet users to share their opinions with others. Previous research has identified privacy and security as two main factors that influence online behavior, especially because of their close relationship to trust (Gefen 2000; Kim et al. 2008). Privacy concerns also play an important role when Internet users decide to sign up for social networks (Fogel and Nehmad 2009). Privacy refers to people's control over the collection and usage of their personal information. Security breaches, meanwhile, include incidents such as hacker attacks or other data thefts. For instance, Kelly (2013) reports that over 250,000 Twitter accounts have been hacked, while Silveira (2012) notes a similar breach for 6.5 million LinkedIn accounts. These examples highlight how personal information might be stolen from Social Media platforms; at the same time, researchers or investors who aim to use Social Media for predicting stock returns need a continuous access to public data, which is generated by Internet users. It is therefore necessary to study consumer behavior with regard to privacy or security breaches.

This dissertation will utilize a laboratory experiment to investigate how people react following privacy and security breaches. So far, researchers have mainly focused on second-order effects when assessing the influence of privacy and security breaches. For instance, various attempts have been made in the literature to show capital market's reactions to such breaches using event study methodology (e.g., Acquisti et al. 2006; Cavusoglu et al. 2004). In contrast, this dissertation will focus on first-order effects—namely, the direct consumer reaction to privacy and security breaches.

1.3 Structure of the Dissertation

Overall, the dissertation consists of five chapters which contain four published research articles: The remaining chapters of this dissertation each cover one of the four published research articles. Table 1-1 provides an overview of the four articles including the research goal, study type, dataset and publication status. Chapter 2 discusses article 1, which deals with whether market anomalies that have been observed in financial markets also occur on Internet platforms. Chapter 3 (article 2) and chapter 4 (article 3) investigate the predictive value of user-generated content with respect to the prediction of share prices; this is done using data from two well-known Social Media applications. Afterwards, chapter 5 (article 4) examines the influence of privacy and security incidents on consumer and investment behavior. In the following, each research article is briefly summarized.

Article 1:   Market Anomalies on Two-Sided Auction Platforms

This article investigates the persistence of market anomalies on an E-Commerce platform. Market anomalies are price distortions that are frequently observed in financial markets. The paper extends this stream of research to the Internet. Anomalies contradict the efficient market hypothesis and can be generally classified as calendar, technical, or fundamental. For the empirical analysis, the authors collected 78,068 transactions between buyers and sellers on a German auction platform between 2005 and 2009. On the platform, sellers offer various products to buyers, such as consumer electronics, jewellery, or cosmetics.

One central finding of the study is that auction platforms are similar to financial markets in that both feature market anomalies. On the platform, there is a persistent turn-of-the month as well as day-of-the-week effect. For instance, prices are lower on Sundays, which might spur buyers to shift their activities to the weekend. Furthermore, prices for the same product are 3.7 percent lower in November compared to January. These effects are statistically significant, but rather small in magnitude. In sum, the results of the study show that auction platforms on the Internet contain marginal market inefficiencies.

Article 2:   Are Crowds on the Internet Wiser than Experts? –
The Case of a Stock Prediction Community

The Wisdom of Crowds describes the phenomenon where large and diverse groups of people can perform better than a few individuals. The power of the WoC has been observed in many different tasks in the offline world. Taking a stock prediction community as an example, this paper contributes a better understanding of the WoC on the Internet. Data was collected from Sharewise, one of the largest stock prediction communities in Europe. Members of Sharewise discuss stocks and assign buy and sell ratings according to their market expectations. This structure makes it possible to aggregate and analyze a large number of predictions. The crowd is diverse to the extent that there are no access limitations and virtually anyone can register on the platform. Therefore, crowd members differ with respect to knowledge, culture, age, gender, and education.

The study aims to compare the stock predictions of these dedicated amateurs with those of professional analysts (who are also present on the platform) as well as the DAX index. Overall, the authors collected 10,146 share recommendations from the crowd between May 2007 and August 2011. The results show that crowd members are able to achieve a return that is, on average, 0.59 percentage points per year higher compared to the professional analysts of banks and research companies. Furthermore, both the crowd members and the professionals outperformed the DAX index during the four-year period.

The platform provider has made several changes over the years, which allowed the authors to study the effects of diversity and independence. The results show that the daily return of the crowd's share recommendations do improve when decisions are made independent from each other. However, there is no significant influence of diversity on daily returns. In sum, the article contributes a better understanding of how diversity and independence influence the performance of an Internet crowd.

Article 3:   **Using Twitter to Predict the Stock Market: Where is the Mood Effect?**

Various studies in finance and psychology have shown that stock markets can be driven by their participants' mood states. This relationship has been shown in laboratory experiments and with a number of different external factors (e.g., weather, time changeover, sport results, air pollution), all of which influence mood states and in turn the stock market (e.g., Hirshleifer and Shumway 2003; Kamstra et al. 2000). While still valuable, these previous studies are limited by their artificiality. In contrast, today's Social Media applications allow researchers to measure mood states in real-time. A few studies (e.g., Bollen et al. 2010) have already indicated that mood states extracted from Twitter and other platforms might have predictive value with regard to stock market developments. However-er, it is unclear whether and to what extent emotional contagion might influence the quality of the results. Several recent articles (e.g., Coviello et al. 2014; Guillory et al. 2011; Kramer et al. 2014) show that emotions on the Internet can spread due to the text-based communication facilitated by Social Media plat-forms.

The article extends this stream of research with a unique sample of roughly 100 million German tweets that were collected between 2011 and 2013. The study also integrates the number of Twitter followers into the analysis, thus providing a first-time investigation into the influence of mood contagion on stock returns. The results reveal that follower-weighted mood states have predic-tive value for stock returns. In cases of improving mood, the DAX index increas-es by 3.3 basis points on the next trading day. In contrast, there is no evidence that aggregate Twitter mood states alone (i.e., without follower numbers) can predict stock returns. These results might be the first indication that emotional contagion on the Internet can influence offline behavior, in particular the will-ingness to invest in stocks.

Article 4:   **The Economic Impact of Privacy Violations and Security Breaches – A Laboratory Experiment**

Several authors have identified data protection as a key factor in the acceptance of Internet platforms (e.g., Fogel and Nehmad 2009). Privacy and security breaches lower consumers' trust and are therefore a serious threat to companies' success. Previous research has primarily focused on capital market reactions when studying the influence of privacy and security breaches. Typically, re-searchers observe decreasing share prices after the corresponding events (e.g., Cavusoglu et al. 2004). In contrast to these studies and their emphasis on second-order effects, this article aims to show the direct consumer reaction via first-order effects. To this end, the authors conducted a laboratory experiment on a university campus. In the experiment, students invested their own money into an investment product being offered by a fictional bank. There were two treatment groups, one of which was informed of a privacy breach while the other was informed of a security breach of the bank in the recent past; there was also one control group, which was not given any additional information.

In this way, the authors were able to measure the influence of privacy and security on both trust and investment amount, which together reflect consumers' willingness to interact with the bank.

According to the results of this study, privacy is very important for building consumers' trust in the company. However, individuals value security more when considering the actual investment decision. The study therefore has implications for practitioners and researchers who want to use Social Media applications for forecasting purposes.

Table 1-1: Dissertation Articles

Con-text	Market efficiency	Wisdom of Crowds	Mood Analysis	Privacy & Security
Study Type and Number of Observations	Field study N = 78,068	Field study N = 10,146	Field study N ~ 100 mio	Laboratory experiment N = 118
Research Goal	Investigating the persistence of market anomalies on an E-Commerce platform	Comparing the accuracy of share recommendations between the crowd and professional analysts	Studying the influence of diversity and independ-ence on the performance of the crowd	Studying the relationship between mood states derived from Twitter and stock returns	Analyzing emotional contagion effects by considering the number of Twitter followers	Investigating the impact of privacy violations and security breaches on the subjects' trust and behavior

Context	Market efficiency	Wisdom of Crowds	Mood Analysis	Privacy & Security
Article	1. Market Anomalies on Two-Sided Auction Platforms	2. Are Crowds on the Internet Wiser than Experts? – The Case of a Stock Prediction Community	3. Using Twitter to Predict the Stock Market: Where is the Mood Effect?	4. The Economic Impact of Privacy Violations and Security Breaches – A Laboratory Experiment
Publication Status	*Presented at: European Conference on Information Systems (ECIS), Barcelona, 2012. VHB Ranking B.*	*Published in: Journal of Business Economics 84(3):303-338. VHB Ranking B.*	*Forthcoming in: Business & Information Systems Engineering. VHB Ranking B.*	*Forthcoming in: Business & Information Systems Engineering. VHB Ranking B.*
Data	Transactions between buyers and sellers on a German auction platform	Share recommendations of Internet users (=crowd) and professional analysts on stock prediction community Sharewise	Public tweets collected through Twitter API	Collected data from university campus
Article	1. Market Anomalies on Two-Sided Auction Platforms	2. Are Crowds on the Internet Wiser than Experts? – The Case of a Stock Prediction Community	3. Using Twitter to Predict the Stock Market: Where is the Mood Effect?	4. The Economic Impact of Privacy Violations and Security Breaches – A Laboratory Experiment

2 Market Anomalies on Two-Sided Auction Platforms[1]

Abstract

A market anomaly (or market inefficiency) is a price distortion typically on a financial market that seems to contradict the efficient-market hypothesis. Such anomalies could be calendar, technical or fundamental related and have been shown empirically in a number of settings for financial markets. This paper extends this stream of research to two-sided auction platforms in Electronic Commerce and empirically analyzes whether calendar anomalies are persistent on such markets. Our empirical study analyzes 78,068 transactions completed between buyers and sellers on a German auction platform and covers the period between April 2005 and May 2009. We observe a persistent turn-of-the-month effect and a day-of-the-week effect that would allow buyers to realize small additional surpluses (0.3 percent price discount). Prices are also persistently lower in the highly competitive Christmas trade period while sellers benefit from higher prices at the beginning of every year. Overall our results support the common notion that two-sided auction platforms are rather efficient markets on which we however can observe some marginal market inefficiencies.

2.1 Introduction

One of the key economic processes when a buyer and a seller engage in trading is that of price discovery, i.e. finding a price that both market sides accept. Due to the characteristics of the Internet, this process has undergone drastic changes over the past few years. Lower menu costs, the reduction of processing costs associated with price differentiation, and new possibilities to interact with prospective trading partners online have led to a number of platforms that employ auction-type mechanisms for price discovery (Bapna et al. 2004; Hinz and Spann 2008; Hinz et al. 2011).

These new platforms share some important properties with financial stock markets but are compared to financial markets still under-researched. In finance,

---

[1] Nofer, Michael / Hinz, Oliver (2012). Market Anomalies on Two-Sided Auction Platforms. European Conference on Information Systems (ECIS), Barcelona, Spain.

the efficient-market hypothesis asserts that financial markets are "informationally efficient". This means that investors cannot consistently achieve higher returns than the average market for a particular risk-class, given that the information is publicly available at the time of the investment (Fama 1970). However, it has been shown empirically that this hypothesis does not hold for markets like stock exchanges and market anomalies occur (e.g., Ariel 1990). A market anomaly (or market inefficiency) is a price distortion on a financial market that seems to contradict the efficient-market hypothesis. Such anomalies could be calendar, technical or fundamental related.

Stock exchanges and auction markets in Electronic Commerce have in common that a market mechanism brings together two market sides, namely buyers and sellers. This could be double auctions or English auctions or any type of auction where prices reflect the relationship between demand and supply. In an efficient market neither buyers nor sellers would be able to systematically gain higher surpluses than market averages for a longer time. If prices are lower for example on Mondays, supply would drop (i.e. sellers are not willing to sell their products/share on this day of the week and offer it on another day of the week) while demand would raise at the same time (i.e. buyers would try to buy on Mondays and thus competition amongst buyers would raise the price) until this distortion is nullified. Thus, if price distortions occur, market forces are expected to correct this distortion so that they cannot persist for a longer time or occur regularly.

If price distortions persist for a longer time, this would allow well-informed sellers and buyers to optimize their market entry and offering/bidding strategy. The intermediary would then have to reconsider his market design and could introduce tools that attenuate information asymmetries to increase the market efficiency. The aim of our paper is therefore to analyze a long time series of data of a two-sided auction platform with respect to market anomalies. We analyze 78,066 transactions in 211 weeks and check whether systematic price and sales distortions occur for products sold on a two-sided auction platform.

The remainder of our paper proceeds as follows: In the following section we will outline the previous research and will focus on research on two-sided markets and introduce the efficient-market hypothesis. We will further outline the empirical research on market anomalies. Section 2.2 will also outline that there are only few studies that examine two-sided auction markets with respect to market anomalies and thereby emphasizes the gap in literature which we will close with our empirical study which is introduced in section 3. We start with describing the platform and its business model before we report some descriptives of the analyzed data. At the core of this section we will analyze whether market anomalies can be found at this platform. We discuss the results in section

4, outline the limitations of our approach and describe opportunities for future research before we conclude the paper with final remarks.

2.2 Previous Research

2.2.1 Two-Sided Markets

A two-sided electronic market is defined as an interorganizational information system through which two customer populations, buyers and sellers, interact to accomplish market-making activities. It helps these customer populations to identify potential trading partners, selecting a specific trading partner and executing the transactions (Choudhury et al. 1998). In two-sided markets, an intermediary provides the platform for linking together two distinct customer populations (Rochet and Tirole 2003). For instance, the auction platform eBay provides the infrastructure as well as the rules and processes to enable transactions between two customer populations: On one side of the market eBay serves sellers with a platform to offer their products, on the other side eBay provides buyers an opportunity to purchase products. Two-sided markets have become more prevalent in the Network Economy (Shapiro and Varian 1998) and can be found in many industries (Eisenmann et al. 2006). The Internet has created new industries such as online auction houses and digital marketplaces (Ellison and Ellison 2005), where intermediaries provide a platform that brings together buyers and sellers or, generally speaking, demand and supply. In 2011, eBay reported to make 60 billion EUR in gross merchandise volume via its E-Commerce platforms (eBay 2012).

In two-sided markets, both customer populations – in case of eBay buyers and sellers interacting on the platform – are crucial to the intermediary. The existence of many sellers offering products on eBay attracts more buyers to the platform. Vice versa, many buyers in turn attract more sellers (Tucker and Zhang 2010). Thus, network effects are present in two-sided markets. Network effects or network externalities are defined as a change in the surplus that a consumer derives from a good or service when the number of consumers or the demand changes (Liebowitz and Margolis 1994).

The rise and fall of several auction and shopping market places has demonstrated the strength of networks effects in this business. Shapiro and Varian (1998) indicate that strong network externalities may lead to a "winner-takes-it-all-market" where one company offers the dominant market place. Moreover, much effort has to be put into the recruitment of new sellers and buyers. On platforms the demand on one side would tend to vanish if there was no demand on the other. Evans (2003) gives a good overview on solutions existing for this

"chicken-and-egg" problem. The literature on competition in two-sided platforms, especially in microeconomics, is growing rapidly, see amongst others
Caillaud and Jullien (2001), Rochet and Tirole (2003) and Rysman (2004).

Market places in the Internet have already reached a mature stage. Late followers can face enormous competitive disadvantages, requiring more marketing
to overcome the barriers-of-entry erected by earlier companies with regard to
consumer preference and awareness (Kerin et al. 1992). Especially in the Internet
late followers suffer from these disadvantages. Particularly the number of electronic market places seems to be limited. Early entrants do have significant advantages and gain large market shares (Hidding and Williams 2003).

However, these electronic markets increase economic efficiency and offer a
high transparency and low search costs for market participants (Bakos 1998) and
thus fulfil better the conditions for perfect markets than traditional offline markets. Among these conditions are perfect market information, no participant with
market power to dictate prices, and no barriers for participants to enter or exit the
markets.

2.2.2   Efficient Markets and Market Anomalies

The efficient-market hypothesis requires that market participants maximize their
utility and have rational expectations. Further, it requires that on average the
market participants are correct. This could also mean that no market participant
is correct, but on average the behaviour of all market participants evens out.
Additionally the market hypothesis requires that market participants update their
beliefs whenever new relevant information becomes available. The market participants do not need to be rational but their behaviour follows a normal distribution. So some market participants might bid too high or too early and some might
bid too low or too late. This yields market prices that cannot be exploited with
certainty to realize abnormal surpluses, especially when considering transaction
and search costs. In such markets, no market participant is better than the rest in
the long run and no market participant has to be right about the market. There are
three common forms of the efficient-market hypothesis: the weak-form efficiency, the semi-strong-form efficiency, and the strong-form efficiency. Each has
different implications for how markets work.

The weak-form efficiency says that future prices cannot be forecasted by
analyzing data from the past. Market participants cannot sustainably realize abnormal surpluses even with access to the entire data of past prices since there are
no patterns of price fluctuations that can be exploited with investment strategies.
This means that prices follow a random walk, but many studies have shown
empirically that markets follow trends for some time. This trend vanishes with
time but may provide excess returns for a short period.

The semi-strong-form efficiency implies that prices incorporate publicly available new information instantly and in an unbiased fashion, such that market participants cannot realize abnormal surpluses by trading on that information. The semi-strong-form efficiency also implies that neither fundamental analysis (since all information available is always priced in) nor technical analysis (since prices follow a random walk) can help to generate abnormal surpluses in the long run. Prices will adapt to new information if the news were previously unknown and relevant. Such news lead to steep upward or downward changes in prices.

The strong-form efficiency implies that prices always reflect all public and private available information and no market participant can generate abnormal surpluses when insider trades are prohibited by law. If the strong-form efficiency is assumed then no fund manager is able to "beat the market" in the long run. Successful fund managers that were able to generate excess returns have thus just been lucky.

We expect that two-sided auction markets would at least be efficient in the weak-form. For used products, the seller has information advantages which lead to information asymmetries (see Akerlof 1970). For boxed, unused products all information is available to all market participants such that even a stronger form could be assumed. On efficient markets (in all forms) market anomalies should not occur according to theory. A market anomaly (or market inefficiency) is a price distortion on a financial market that seems to contradict the efficient-market hypothesis. Such anomalies could be calendar, technical or fundamental related and have been shown empirically in a number of settings for financial markets.

One prominent example for a calendar anomaly is the Monday effect which manifests in the belief that securities market returns on Mondays are on average less than other days of the week, and are often negative on average. The Monday effect, which is also known as Day-of-the-week effect or weekend effect, has been observed in both American and foreign exchanges (e.g., Fields 1931; Jaffe and Westerfield 1985). Ariel (1987) and Lakonishok and Smidt (1988) observed the tendency of stock prices to increase during the last two days and the first three days of each month. This effect, which is called Turn-of-the-Month effect, is most likely based on the timing of monthly cash flows of pension funds that invest in turn at this time of the month in the stock market. Lakonishok and Smidt (1988) also observed another calendar related market anomaly which they called holiday effect. Their empirical study revealed that investors can generate abnormal returns before an exchange-mandated long weekend or holiday such as Labor Day or Christmas.

Fundamental related anomalies are for example the small-cap effect (e.g., Roll, 1981), which describes the tendency that small-capitalization stocks outperform the market or the value effect (e.g., Fama and French 1998), which refers to the positive relation between security returns and the ratio of accounting based measures of cash flow or value to the market price of the security. These fundamental related and technical related anomalies, like the momentum effect (see Jegadeesh and Titman 1993), exist exclusively in financial markets and are therefore out of the scope of this paper while calendar anomalies can in principle also exist on other markets like two-sided auction markets.

Yet, only few studies examine market anomalies in two-sided markets in Electronic Commerce. Brynjolfsson and Smith (2000) showed for example that the price dispersion for the same product on electronic markets is still substantial and thus the rule of a single market price is invalid. Ba and Pavlou (2002) revealed that reputation has a significant influence on prices in electronic markets (see Dellarocas 2003 for another overview). This finding is still in line with the efficient-market hypothesis, since a more reliable trading partner can lower the risk of being cheated by the trading partner.

On perfect markets demand should follow supply and vice versa. If demand is higher on certain days on auction platforms like eBay, prospective sellers would shift their offers to these days since they would expect higher prices and supply would rise equally. It would thus not be possible to make additional profits for sellers or achieve lower prices for buyers on particular weekdays or months for a longer period of time. However, first empirical evidence suggests that two-sided platforms suffer from persistent market anomalies that substantially influence sales and prices.

Simonsohn (2010) for example shows that a disproportionate share of auctions end during peak bidding hours and such hours exhibit lower selling rates and prices. Moreover, Simonsohn (2010) also finds that peak listing is more prevalent among sellers likely to have chosen ending time strategically, suggesting disproportionate entry is a mistake driven by bounded rationality rather than mindlessness.

This paper picks up the topic and continues to empirically examine market anomalies in two-sided electronic markets, i.e. auction platforms in particular, and focuses on calendar anomalies.

2.3 Empirical Study

For the purpose of our study, we acquired data from a German auction platform on a daily basis and analyze the sales, price, and revenues of the top-selling products over time. We label this intermediary "Platform.com", since we do not disclose the name for reasons of maintaining confidentiality.

2.3.1 Platform Description

On Platform.com, sellers offer their products – such as consumer electronics, household appliances, jewellery, watches and cosmetics – to buyers. The sellers are exclusively professional retailers (in contrast to eBay) who sell to private individuals. Only brand-new, boxed products can be sold on Platform.com. The European article number (EAN) is used to identify the products unambiguously.

The intermediary is a start-up company funded in three rounds, and investors include the High-Tech Entrepreneur Fund of the German Federal Ministry of Economics and Technology. Platform.com charges sellers per transaction, while buyers can use the platform for free. Sellers pay 3 percent fee on the volume per transaction while the intermediary does not charge buyers. One market side thus subsidizes the customer population on the other market side which is a common feature in the network economy and two-sided markets. The relationship between the intermediary and its customer populations, buyers and sellers, is non-contractual. Prices include shipping costs and the intermediary offers a trusted service and takes responsibility for the sellers' action on the platform such that buyers cannot suffer from auction fraud. Therefore, the intermediary chooses the sellers quite carefully since the intermediary has to cover potential losses due to sellers' fraudulent actions.

The intermediary applies a continuous double auction to find prices which makes the platform thus comparable to common stock exchanges. In continuous double auctions multiple buyers and multiple sellers can simultaneously and continuously negotiate for the same type of good. The bids comprise offers to buy or offers to sell. Each incoming bid is matched with the best possible order on the opposite market side. If the incoming bid matches an open offer on the other market side, the intermediary initiates the transaction, otherwise the bid is collected in the order book as open. In the case of Platform.com, all prospective market participants have access to the order book and can thereby evaluate the market situation.

2.3.2 Descriptives

Our study comprises transaction data between buyers and sellers on Platform. com and covers the period between April 2005 and May 2009. The prices range between 0.70 EUR and 4,199.00 EUR with a mean price of 106.18 EUR. Overall, 351 different sellers sold 25.677 unique products types (as identified by the unique EAN) in 78,068 transactions to 65,894 different buyers. As this numbers indicate, the retention rate for sellers is quite high while the retention rate for buyers is very low. Most buyers only buy one product on Platform.com which the intermediary certainly has to improve if it wants to capture a significant market share in the auction market.

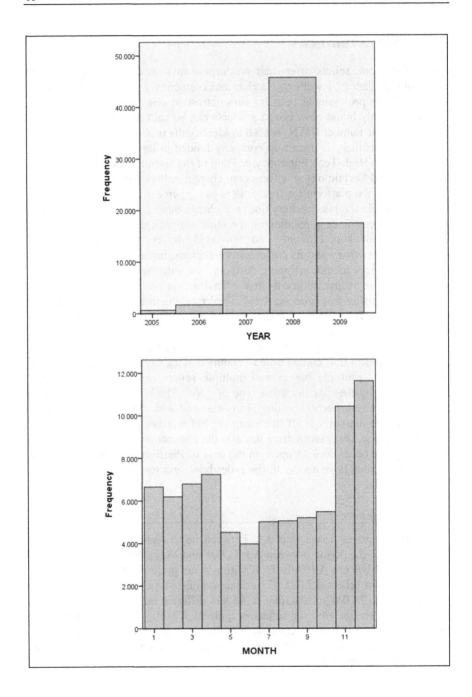

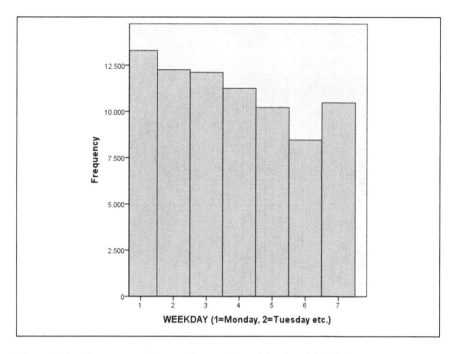

Figure 2-1: Frequency of Transactions per Year, Month and Weekday

We also plot the frequency of sales for the days of the month (see Figure 2-2). We find that at the first days of the month the number of transactions increases. The number of sales then decreases until the 25th of the month and then starts to increase again. Note that the number of transactions on the 29th, 30th and 31st is biased downwards because not every month has a 31st day etc. Cash flows (wages, rent received etc.) typically occur at these days of the month while the account balance typically drops over the month.

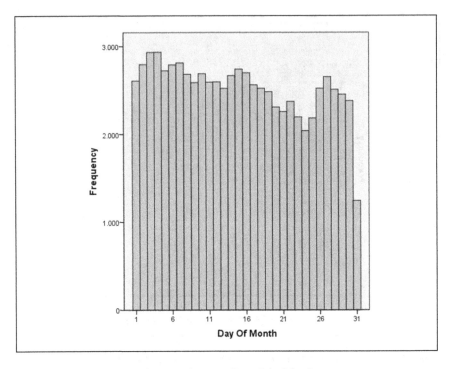

Figure 2-2:   Frequency of Transactions per Day of the Month

The results presented in Figure 2-1 and Figure 2-1 indicate that demand and supply fluctuate over time but these first results do not allow to draw conclusions about the efficiency of two-sided auction markets and this platform in particular. We therefore analyze the effect of calendar regularities on sales and price quantitatively in the next chapter.

2.3.3   Analysis

We start to analyze the effect of seasonality on daily sales and thus aggregate the transactions on a daily basis. We use dummy variables for weekdays (e.g., Monday = 1 indicates that the observation was made on Monday, 0 = otherwise), for months and for years to account for the development over time.

Additionally, we introduce a dummy variable for public holidays (1 = for all public holidays in Germany, 0 = otherwise). We further create a dummy variable for the first five days and the last four days of each month and call the variable "Turn of the month". We use Monday, December and the year 2009 as reference and thus estimated coefficients reflect changes in contrast to these reference points.

To account for autocorrelation (Durbin-Watson statistic indicated the presence of autocorrelation in the residuals), we estimate a Prais-Winsten generalized least-squares regression, which assumes AR(1) error structures but no dynamics and which is equivalent to the use of full-information maximum likelihood for an AR(1) model. Additionally, we use robust standard errors to account for heteroskedasticity (White-Test $p < 0.01$). Table 2-1 shows the results of the Prais-Winsten regression with robust standard errors. The F-Test indicates that the model is valid and the transformed Durbin-Watson statistic is also in a good range. The variance inflation factors (VIF) are below 4 (mean VIF = 1.96) and thus multicollinearity does not seem to be a problem in the estimated model. The model estimated explains 24 percent of the variance and provides face validity.

We do not observe a significant effect of the turn of the month ($p > 0.3$) on sales on our examined auction platform. We do however observe that sales on all days of the week are significantly lower than on Mondays ($p < 0.01$). There are about 24 sales less on Saturdays than on Mondays. We also see significantly fewer sales in the years 2005-2008 than in 2009. The results further illustrate a strong seasonality over the year. In the first half of the year on average about 50 less sales occur per day while the model reveals a brisk Christmas trade period. In November and December sales peak out which also Figure 2-1 illustrated. On public holidays sales drop by about 4 ($p<0.05$).

These figures confirm a strong seasonality and a periodic fluctuation of sales. The question however remains whether both market sides adapt in the same speed such that no price distortions occur. In a perfect market we should not observe persistent market anomalies. We will examine this interesting question in a next step and estimate a model with price as dependent variable.

To make the observations comparable, we standardize the price of each observation by dividing by the mean price for this product. Again, we apply a Prais-Winsten generalized least-squares regression with robust standard errors and follow the same approach as outlined above. Table 2-2 depicts the results in detail.

The F-Test statistic allows to reject the null hypothesis that the set of instruments are jointly zero, all VIFs are below 3 (mean VIF = 1.55), the model provides face validity, and explains 16 percent of the variance. The constant of 97.6 percent is highly significant. We observe that prices around the turn of the month are slightly (-0.1 percent, $p < 0.1$) lower than during the rest of the month which is a first evidence of market anomalies. We also observe that in the observation period of more than four years, prices are significantly lower on Sundays than on Mondays ($p < 0.05$). The same product generates on average 0.3 percent

Table 2-1:   The Effect of Seasonality on Number of Daily Sales

	Coef.	Semirobust Std. Err.	t	P>|t|
Constant (***)	209.80	21.52	9.75	0.00
Turn of the Month	-2.06	2.01	-1.02	0.31
Tuesday (***)	-5.46	1.19	-4.57	0.00
Wednesday (***)	-5.72	1.59	-3.61	0.00
Thursday (***)	-9.59	1.73	-5.55	0.00
Friday (***)	-14.26	1.83	-7.80	0.00
Saturday (***)	-23.68	1.92	-12.35	0.00
Sunday (***)	-13.55	1.78	-7.60	0.00
Year2005 (***)	-172.71	18.99	-9.09	0.00
Year2006 (***)	-152.73	18.03	-8.47	0.00
Year2007 (***)	-120.50	16.89	-7.14	0.00
Year2008 (***)	-38.85	15.55	-2.50	0.01
January (***)	-57.95	10.78	-5.37	0.00
February (***)	-59.64	12.01	-4.97	0.00
March (***)	-56.73	12.43	-4.56	0.00
April (***)	-57.83	12.97	-4.46	0.00
May (***)	-53.60	13.02	-4.12	0.00
June (***)	-47.34	12.76	-3.71	0.00
July (***)	-41.88	12.78	-3.28	0.00
August (***)	-49.10	14.22	-3.45	0.00
September (*)	-23.26	14.13	-1.65	0.10
October (*)	-20.65	12.56	-1.64	0.10
November	-14.38	11.48	-1.25	0.21
Public Holiday (**)	-4.21	2.05	-2.05	0.04
*** $p<0.01$; ** $p<0.05$; * $p<0.1$; $F = 14.65$ $(p<0.01)$; $R^2 = 0.24$				

less revenue on a Sunday than on the next day. According to the platform operator, the number of prospective buyers is lower on Sundays while most of the sellers keep their offer open for the weekend. This leads to a systematic excess supply. The incentive of 0.3 percent lower prices is however not enough to trigger a sufficient number of buyers to go online on Sundays which would close the gap in a perfect market. On average, the price discount on Sundays is about 0.32 EUR in absolute terms. The results also reveal that prices drop on average over time.

This can be caused by dropping prices over time (e.g., pricing over normal life cycle) or a growing competition amongst sellers. Over the year, sellers can realize the highest price premium during the first half of the year. Prices are 1-2 percent higher in this period than in December. This can result in savings of around 2.80 EUR for an average priced product.

From a prospective buyer's perspective, the best prices can be realized in November with a discount of 1.1 percent compared to December, or a discount of 3.7 percent when compared to January. This phenomenon is persistent and indicates that the market is either not informationally efficient or market participants behave strategically. Based on our input from management, we conjecture that the second alternative seems plausible. In the Christmas trade period a lot of sellers try to capture market share by acquiring buyers on the platform and turn them in loyal customers after this period of hard competition.

2.4 Discussion

Our analyses revealed that two-sided auction markets like Platform.com show the same properties as financial markets. Activity fluctuates over the week, over the year and over time. Interestingly, demand and supply increase/decrease quite simultaneously so that prices are rather constant over time. We however observe small market anomalies and persistent price distortions over time. Although we examined a period of more than four years, market participants can realize additional surpluses by shifting demand/supply to off-peak periods. We can predict that prices are on average lower on Sundays so that prospective buyers should move their demand to this day of the week. We also observe that there is a fierce competition among sellers before Christmas. Buyers can benefit from this competition. Product prices for the same product differ by 3.7 percent between November and January.

From a tactical point of view, sellers should revisit their activities in the Christmas trade period as supply increases more than demand. Sellers can realize higher prices after Christmas. A price premium of 3-4 percent is huge, especially for low-margin consumer electronics that are mainly sold on Platform.com. However, our analysis does not consider strategic behaviour like the acquisition of new customers and the build-up of a long customer relationship. Selling on the platform in the Christmas trade period might lead to new customers with a high customer lifetime value. Therefore, the Christmas trade period might be considered as investment that will pay off in the long run. We also observe that price premiums decrease over time which indicates a growing competition and decreasing prices over the lifetime of the products.

Although demand and supply fluctuates heavily over time (see Figure 2-1), prices stay relatively constant. We observe persistent price distortions over time but their magnitude is rather small in comparison to the fluctuation of sales. This

Table 2-2:   The Effect of Seasonality on Standardized Prices

	Coef.	Semirobust Std. Err.	t	P>|t|
Constant (***)	0.976	0.002	601.58	0.00
Turn of the Month (*)	-0.001	0.001	-1.87	0.10
Tuesday	0.001	0.001	1.22	0.22
Wednesday	0.001	0.001	1.18	0.24
Thursday	0.000	0.001	0.28	0.78
Friday	0.001	0.001	0.82	0.41
Saturday	-0.000	0.001	-0.09	0.93
Sunday (**)	-0.003	0.001	-2.38	0.02
Year2005 (***)	0.028	0.005	5.29	0.00
Year2006 (***)	0.043	0.003	12.85	0.00
Year2007 (***)	0.048	0.002	27.64	0.00
Year2008 (***)	0.017	0.001	15.02	0.00
January (***)	0.026	0.002	13.97	0.00
February (***)	0.022	0.002	13.25	0.00
March (***)	0.020	0.002	12.21	0.00
April (***)	0.020	0.002	12.48	0.00
May (***)	0.012	0.002	7.15	0.00
June (***)	0.006	0.002	3.45	0.00
July	0.001	0.002	0.71	0.48
August (***)	-0.005	0.001	-3.32	0.00
September (***)	-0.007	0.001	-4.45	0.00
October (***)	-0.008	0.001	-5.57	0.00
November (***)	-0.011	0.001	-8.61	0.00
Public Holiday	-0.001	0.002	-0.12	0.91
*** $p<0.01$; ** $p<0.05$; * $p<0.1$; $F = 88.16$ ($p<0.01$); $R^2 = 0.16$				

clearly indicates that the market is rather efficient and supply follows demand and vice versa. This is especially interesting since Platform.com does not provide tools that allow analyses of historical data such that both market sides can suffer from incomplete information. Obviously market forces shift the market partici-pants in the right direction and the market is thus quite efficient. The market anomalies are rather small in magnitude and would not allow to carry out arbi-trage strategies. Based on our analyses, we can also conclude that 0.3 percent price discount is obviously not enough for some prospective buyers to go online on Sundays. The average saving of 0.38 EUR does not compensate the oppor-tunity costs.

2.4.1   Limitations and Future Research

Our analysis is obviously restricted to the single case of Platform.com. The generalizability of our results must therefore be revalidated with data from additional two-sided auction platforms (e.g., eBay). With the access to the entire dataset of transactions conducted on Platform.com we are able to examine a very long period and a high number of products. This is certainly a unique selling proposition of this analysis. However, we did not separate pioneer buyers (2005-2006) from followers (2007-2009) so that our results might be influenced by changes in the customer population which would be an interesting avenue for future research. Based on the available data we can however not assess the intention of the market participants. Selling in the Christmas trade period might make sense in the long run although prices are systematically lower in this time of the year.

Overall, there could be many reasons for differences in the demand. For instance, it is possible that people buy less on Sundays since they pursue their hobbies or spend time with their families. However, our data does not provide any further information about the buyers' preferences which is why we can only speculate at this point. Future research might take these reasons for differences in the demand into account.

We find that the magnitude of the market anomalies is rather small and it would be interesting to compare this magnitude of market anomalies to one-sided auction markets where auctioneer and seller are the same person. In the case of one-sided auction platforms, the seller could predict demand based on historical data as well as market research and would be able to perfectly adapt supply to the forecasted demand. This comparison would allow a more meaningful evaluation of the efficiency of two-sided markets where both market sides have only access to incomplete information. Our analysis does not incorporate competitive actions which is another limitation. We expect that data on exogenous events like marketing activities of other auction platforms would help to explain more variance. Overall, a more sophisticated model that also captures the diffusion and growth process of the platform as well as endogenous factors like investments in the functionality of the platform would help to explain a higher fraction of variance. We are however confident that our model does not systematically bias estimates and that our findings are robust.

2.4.2   Conclusion

While researchers in finance have intensively investigated the efficiency of financial markets, the efficiency of two-sided auction markets in Electronic Commerce still remains an open question. Our work contributes to research in this area by analyzing transaction data for a period of more than four years. We find that sales fluctuate heavily and follow a predictable pattern. The differences in sales are huge on the examined auction platform. The differences in prices are however small which indicates that supply and demand simultaneously increase and decrease. We observe however calendar related market anomalies that are persistent over time. Prices are lower on Sundays and on the turn of the month. These differences are small but significant and an evidence of market anomalies from an academic point of view.

Prices change also over the months and provide practitioners to optimize their bidding and selling strategy. Buyers benefit from a brisk Christmas trade period with high competition while sellers can generate excess returns from focusing on the weeks after Christmas. We conclude that research on two-sided auction markets can further benefit from empirical analyses of transactional data. Unfortunately these data are not as publically available as data on financial markets.

3 Are Crowds on the Internet Wiser than Experts? – The Case of a Stock Prediction Community[1]

Abstract

According to the "Wisdom of Crowds" phenomenon, a large crowd can perform better than smaller groups or few individuals. This article investigates the performance of share recommendations, which have been published by members of a stock prediction community on the Internet. Participants of these online communities publish buy and sell recommendations for shares and try to predict the stock market development. We collected unique field data on 10,146 recommendations that were made between May 2007 and August 2011 on one of the largest European stock prediction communities. Our results reveal that on an annual basis investments based on the recommendations of Internet users achieve a return that is on average 0.59 percentage points higher than investments of professional analysts from banks, brokers and research companies. This means that on average investors are better off by trusting the crowd rather than analysts. We furthermore investigate how the postulated theoretical conditions of diversity and independence influence the performance of a large crowd on the Internet. While independent decisions can substantially improve the performance of the crowd, there is no evidence for the power of diversity in our data.

3.1 Introduction

Since popular science author James Surowiecki published his seminal book about the Wisdom of Crowds (WoC) in 2004, this phenomenon has been increasingly discussed by researchers from various disciplines in recent years (e.g., Hertwig 2012; Koriat 2012; Simmons et al. 2011). According to the WoC, a diverse and independent "crowd" can make more precise predictions than a few people, even when only professionals are involved. In this article we follow Poetz and Schreier (2012) who define the crowd as a "potentially large and unknown population" (pp. 246). While the WoC can be widely explained by math-

---

1 Nofer, Michael / Hinz, Oliver (2014). Are Crowds on the Internet Wiser than Experts? – The Case of a Stock Prediction Community. Journal of Business Economics, 84 (3), 303-338.

ematical principles (Galton 1907; Hogarth 1978; Treynor 1987) it is closely related to the concept of collective intelligence and many authors use these terms as synonyms (Kittur and Kraut 2008; Leimeister et al. 2009; Surowiecki 2004). The emergence of collective intelligence has been observed in many disciplines, and the Internet, with its low communication and processing costs (Schwind et al. 2008), may especially foster this phenomenon. The Internet is particularly suitable for studying the conditions for the WoC phenomenon since diverse people from different places in the world interact with each other on websites, blogs and message boards. The rise of social media applications over the last decade fosters the extensive communication process among Internet users.

One often-cited example for this collective intelligence is the online encyclopedia Wikipedia. The accuracy of Wikipedia's science entries – written collectively by Internet users – was found to be virtually as good as Britannica's articles. Thus, an unpaid crowd did match a few professional editors (Giles 2005). The high quality of Wikipedia articles was recently confirmed by Rajagopalan et al. (2011) who compared cancer information between Wikipedia and a professional database.

The strength of the Internet also becomes apparent when it is used to predict future events. Researchers successfully forecasted flu epidemics with Google (Ginsberg et al. 2009), music sales with blogs and social networking sites (Dhar and Chang 2009), and election winners or movie sales with prediction markets (Berg et al. 1997; Forsythe et al. 1999; Spann and Skiera 2003).

This promising evidence prompted companies to utilize the WoC for business purposes. The concept of open business models is not totally new. Many companies are already "crowdsourcing" (Howe 2008) tasks to a large group of committed people on the Internet and involve the crowd in solving business problems or in developing new products (Jeppesen and Frederiksen 2006; Leimeister et al. 2009). Generally, the integration of customers and Internet users is not restricted to high-technology or software companies. Chesbrough and Crowther (2006) identified many industries (e.g., chemicals, medical devices, aerospace), which successfully use the open innovation concept.

Previous research also suggests that the financial industry might consider the crowd's opinions for their investments although practical evidence is rather scarce. For instance, the predictive value of user-generated content with regard to share returns bas been identified (Antweiler and Frank 2004; Avery et al. 2009; Bollen et al. 2010). Hill and Ready-Campbell (2011) used data from the Motley Fool CAPS, which is a stock prediction community in the USA. The authors show that the members of the community outperform the S&P 500 with their investment decisions. Moreover, using a genetic algorithm, superior investors from the crowd can be identified based on their prior stock recommendations. In a similar study, Avery et al. (2009) conclude that "CAPS participants possess price-relevant information that is far from systematically incorporated in market prices" (pp. 35). However, these studies used a platform where members can only assign buy or sell ratings for stocks and are not able to close recommendations or specify price targets on their own.

The sole information with respect to the superior accuracy of user-generated stock predictions compared to indices is not enough to draw conclusions about the real value for banks and investors since professional analysts might still be more accurate with their decisions. Further, none of the existing studies investigated the drivers of the WoC on the Internet so that forecast accuracy might be improved by a changed setting. So far, authors only investigated preconditions that must be met for a wise crowd in the offline world: members should be knowledgeable, diverse and independent. In addition, participants should be motivated enough (Page 2007; Simmons et al. 2011; Vul and Pashler 2008).

Our research is therefore twofold: First, we deal with the question of whether the crowd is able to make better share price forecasts than professional analysts from banks, brokers and research companies. Second, we aim to compare the forecast accuracy of the crowd under varying degrees of diversity and independence. The article primarily focuses on the comparison between the crowd and experts, while the influence of diversity and independence is our subordinate research objective.

We will close the gap in the literature by taking a stock prediction community as an example. We collected data from one of the largest stock prediction communities in Europe. The platform publishes recommendations of the crowd as well as professional analysts. Our research therefore distinguishes from early approaches in that every recommendation has a price target and can be opened or closed on regular trading days. Thus, we can precisely determine the performance and the duration of the recommendations. Moreover, we can compare the accuracy of the crowd with the accuracy of professional analysts and ultimately examine the prediction accuracy under different conditions.

The reader should always take the perspective of an investor who is seeking advice rather than looking for the best portfolio or trading strategy. Experimental as well as empirical research shows that many investors rely on analyst recommendations (Kelly et al. 2012). This is especially true for smaller investors (Malmendier and Shanthikumar 2007). Our study shows whether this confidence is justified or whether investors should better rely on the Internet crowd.

The remainder of our paper proceeds as follows: In the following section we will first summarize important facts about stock prediction communities before presenting the theoretical background which is necessary to derive our hypotheses. We discuss the value of expertise in forecasting and the conditions that must be met for a wise crowd. Section 3.3 describes the setup of our empirical study while section 3.4 presents the results. We conclude with implications and a summary of our results before pointing out limitations and making suggestions for future research.

3.2 Previous Research

In this section we will first outline was has been found about stock prediction communities and the WoC in general. Afterwards we present the theoretical background and related hypotheses.

3.2.1   Domain Background

Before the emergence of sophisticated stock prediction communities, people discussed the stock market development on the Internet with the help of blogs or message boards (e.g., Yahoo! Finance). Computational linguistics methods allow researchers to determine the quality of Internet posts as well as classify them into positive, negative or neutral opinions about the respective company (Gu et al. 2007). For instance, Antweiler and Frank (2004) analyzed the content of stock message boards and found that the volume of messages can predict market volatility as well as stock returns. However, the authors did not find a statistically significant relationship between positive ("bullish") comments and share price returns.

With the rise of Social Media, stock prediction communities have gained more and more attraction. Prominent examples of stock prediction communities include the Motley Fool CAPS, Piqqem or Covestor. These platforms differ from traditional stock message boards to the extent that community members not only discuss about companies but also pick stocks according to their expectations. Thus, there is no need to interpret positive or negative signals by analyzing text messages. Instead, buy and sell recommendations can be clearly identified depending on whether an Internet user is buying or selling the respective stock.

3.2.2   Theoretical Background

Comparison between Professional Analysts and the Crowd

Although previous research indicates that user-generated content can be used to predict share returns, until now, no study has compared the performance of the crowd with the recommendations of highly paid financial experts in the real world. In our case, we describe the term "expert" as someone who performs a task for working reasons and spends plenty of time with his profession. Thus in our context, an expert is defined as a professional analyst from a bank or research company who has a lot of experience in his area of expertise: publishing share recommendations and predicting the stock market development.

Cognitive-science research attests that experts, at least within their domains, possess superior skills and thinking strategies compared to novices (Anderson 1981; Larkin et al. 1980). However, studies from finance, economics, medicine and other research areas suggest that the value of expertise is limited, especially in terms of forecasting future developments (Johnston and McNeal 1967). For instance, Levy and Ulman (1967) presented subjects, who had varying expertise in psychology (i.e., professional mental health workers, student mental health workers and people with no mental health experience at all), 96 pictures – half painted by psychiatric patients and half by normal people. Accuracy in distinguishing the healthy people from the patients did not depend on the participants' expertise.

Armstrong (1980) developed the seer-sucker theory concluding that "No matter how much evidence exists that seers do not exist, suckers will pay for the existence". According to the author, this simply means that expertise is of little or no value at all. Armstrong refers to many examples in the literature which support the seer-sucker theory. For instance, Taft (1955) shows that professional psychologists are worse in judging people compared to non-psychologists.

On the Internet, you can find evidence suggesting that a large crowd is at least able to keep up with experts. The accuracy of Wikipedia's science entries that are mostly written by dedicated amateurs matches those of the professional Encyclopedia Britannica (Giles 2005). Spann and Skiera (2003) compared the Hollywood Stock Exchange prediction market with expert predictions in terms of new movies' box-office success, and found that the experts could not substantially outperform the crowd on the opening weekend.

With regard to expertise in the financial industry, it is widely known that professional financial analysts and fund managers tend to underperform the broader market with their investment decisions (Carhart 1997; Jensen 1968; Malkiel 1995). Although share analyses exert considerable influence on the market participants' investment decisions through the media, the accuracy of these forecasts has been found to be quite poor (Diefenbach 1972). Bogle (2005) studied two 20-year periods between 1945-1965 and 1983-2003 and found that the average equity fund return fell short of 1.7 percentage points of the S&P 500 return in the first case and 2.7 percentage points in the latter case. Taking another benchmark index, the author previously showed that between 1984 and 1999, about 90 percent of all mutual funds achieved a lower return than the Wilshire 5000 index, which measures the performance of all publicly traded shares in the USA (Bogle 2001).

In contrast to professional share analyses, first studies in the area of stock prediction communities show promising results with respect to the crowd's ability to beat the broader market. For instance, Hill and Ready-Campbell (2011) found that the Internet crowd is able to outperform the S&P 500 by 12.3 percentage points during 2008. Collectively, the findings about the value of expertise suggest:

H1:   Members of a stock prediction community on the Internet (=crowd) are able to achieve a higher daily return than professional analysts (= experts).

Conditions for a Wise Crowd

Researchers from various disciplines are preoccupied with the question why the WoC actually works. According to a wide range of studies, there are four conditions that must be met for a wise crowd: knowledge, motivation, diversity and independence (Simmons et al. 2011).

In our given setting, we assume that the platform leads to a typical self-selection towards knowledgeable and motivated users. We believe that most of the participants have a certain amount of knowledge about the stock market and are motivated enough. Members would not voluntarily register on the platform and spend plenty of time for sharing their opinions with other members if they had little knowledge or motivation. Antweiler and Frank (2004) studied content of Internet stock message boards and refer to theories of DeMarzo et al. (2001) and Cao et al. (2002) in order to explain the motivation for posting messages. For stock market participants it might be profitable to gain influence in the community since recommendations can affect share prices if other investors follow. Participation is also driven by the willingness to learn from other members, especially in the case of sidelined investors.

The degree of independence and diversity is however changing on the platform – thus affording the opportunity of a natural experiment – so that we can examine their effect on the performance of the crowd.

Diversity

The reason why diverse groups are often doing better is grounded in the fact that they are more able to take alternatives into account. A number of studies investigated the problem-solving effectiveness of groups depending on their composition. For instance, Watson et al. (1993) found that in the long run, groups with a higher cultural diversity generate more alternatives and a broader range of perspectives. Hong and Page (2001) present a model showing that diversity in terms of the workers' perspectives significantly enhances their ability to solve even difficult problems.

Organization science has also been reflecting on the optimum composition of working groups (Williams and O'Reilly 1998). Informational diversity is considered as a key driver of performance (Jehn et al. 1999), although too much informational overlap was found to be counterproductive (Aral et al. 2008).

Researchers also focused on demographic characteristics, which are assumed to correlate with expertise and cognitive skills. Bantel and Jackson (1999) showed that innovation in banks is positively influenced by the diversity of their management teams with regard to age, education and functional experience. Kilduff et al. (2000) found a positive relationship between age diversity of top management teams and firm performance. Besides age, Elron (1997) included tenure, functional background and education for measuring heterogeneity of management teams, also observing a positive relationship between cultural diversity and team performance.

Reagans and Zuckerman (2001) found support for the positive relationship between diversity and productivity in the sense that network heterogeneity leads to more communication among team members with different organizational tenure. This reduces demographic boundaries and enables access to different information, perspectives and experiences.

Further evidence for the superior performance of diverse groups comes from March (1991). While homogenous groups that are composed of only long-term employees focus on exploiting the existing knowledge, heterogeneous groups with a mixed composition of employees are better at exploring new ideas and alternatives. Although being less knowledgeable than their experienced senior colleagues, new recruits enhance the diversity and therefore make the entire group smarter regardless of their individual abilities. This is due to the novel information brought to the group.

With respect to financial investment decisions, numerous studies report differences between men and women. Sunden and Surette (1998) found that gender diversity exerts an influence on the asset allocation of retirement savings plans. Other evidence for the different investment behavior with respect to pensions comes from Bajtelsmit and VanDerhei (1996) and Hinz et al. (1997) who show that women invest more conservatively than men. Researchers explain these differences by investigating risk preferences: women tend to be more risk averse than men (Arch 1993; Byrnes et al. 1999; Jianakoplos and Bernasek 1998). These differences can be explained by the level of overconfidence. Research in psychology demonstrates that in general men are overconfident (Deaux and Farris 1977; Lewellen et al. 1977). According to Prince (1993) men also feel more competent with respect to financial decision making. Overconfident investors hold riskier portfolios (Odean 1998) and are more prone to excessive trading which leads to decreasing returns (Barber and Odean 2001).

The overall conclusion from this line of research is that diversity opens possibilities for gaining access to different sources of knowledge and information, which fosters problem solving and overall performance. Further, differences in preferences or opinions among crowd members (e.g., caused by gender differences) ensure that collective errors will be reduced and estimates converge to the correct values. Collectively, the findings about diversity suggest:

H2:   Increased diversity among the members of the crowd will lead to higher
      daily returns of recommended stocks.

Independence

In contrast to diversity, the prevailing view in the literature with respect to the influence of independence is not clear. Independence means that each crowd member can make his or her decision relatively freely and without being influenced by other opinions (Surowiecki 2004). According to previous research, independence is often shown to serve as a positive driver for the performance of groups. By means of a laboratory experiment, Lorenz et al. (2011) revealed that little social influence within the group can be enough to reduce the WoC effect. The subjects in this experiment successively had to answer several estimation questions with regard to geography and crime statistics, and were exposed to different degrees of social influence. Participants either received full, aggregated or no information at all about their group members' estimates. This study revealed that the information about the others' opinion alone leads to a convergence of the answers without improving the accuracy of the decision in terms of collective error.

If individual decisions depend on the previous behavior of others, herding or so-called "information cascades" can result just because of the assumption that the others are better informed. Informational cascades occur when people ignore their private information and blindly follow the crowd. This pattern has been shown theoretically as well as empirically for investment recommendations (Graham 1999; Scharfstein and Stein 1990) and also for general social and economic situations (Banerjee 1992; Bikhchandani et al. 1992; Hinz et al. 2013). In the area of finance, herding means that investors' behavior converges. Welch (2000) has shown that the recommendation revision of a security analyst positively influences the next two revisions of other analysts. Interestingly, the influence of the consensus estimate on the recommendation revision of analysts is not affected by its previous accuracy. Thus, herding can obviously happen without the certainty of correct evaluations, which is why individuals sometimes seem to irrationally rely on other opinions.

However, following the crowd does not always need to be irrational. Scharfstein and Stein (1990) presented a model that assumes herding as rational behavior among investment managers. In the case of a wrong decision, the reputation only suffers if the responsible manager was the only one who bought the bad product. This is why even good managers herd on bad decisions instead of taking the risk to fail exclusively. The theoretical insights from the model have been tested empirically: Graham (1999) found evidence for herding behavior among investment newsletters. The author observed newsletters that herd on the investment advices of the best known and well-respected newsletter "Value Line".

More evidence for behavior adaption comes from Banerjee (1992) and Bikhchandani et al. (1992), who presented models showing that information cascades can arise when people believe that the other persons have superior information. This leads to a loss of private information since individuals adopt the behavior of others instead of relying on their own information. Similarly, Hinz and Spann (2008) found that information coming from strong ties can decrease the performance of economic decisions, while information coming from distant parts of social networks can have a positive impact on said performance.

Despite this broad evidence for the power of independence, there are also notes in the literature which indicate the opposite. Especially research on forecasting provides examples, which show that communication among members can improve the overall group performance. Prediction markets, such as the Iowa Electronic Markets (prediction of election winners) or the Hollywood Stock Exchange (prediction of new movies' box-office success), allow people to trade virtual stocks that receive payoffs depending on the outcome (Wolfers and Zitzewitz 2004). On these platforms, stock prices are visible to all members so that independence is rather small.

Another example where decisions depend on observable opinions of other participants is the Delphi method. Participants are repeatedly asked to answer questionnaires. The fundamental idea is to achieve convergence to the true value by iterating question rounds. After providing their own beliefs, participants receive the opinions of other members as well as arguments for the decision (Dalkey and Helmer 1963). Although both prediction markets as well as Delphi studies violate the condition of independence, the accuracy of these methods has found to be quite high (Ammon 2009; Forsythe et al. 1999; Spann and Skiera 2003).

Despite this evidence from Delphi studies and prediction markets, we expect that stock predictions of an Internet community will benefit from more independence. Financial markets have shown to be particularly vulnerable to herding, information cascades and other effects which are threats to independence. Thus, we hypothesize:

H3:   A higher degree of independence among the members of the crowd will
       increase the daily return of recommended stocks.

3.3 Setup of Empirical Study

3.3.1   Data Collection

We collected data from one of the largest European stock prediction communities on which members can assign buy or sell ratings, enter price targets and
precisely quantify their expectations on the stocks' performances. This website
publishes stock recommendations of dedicated amateurs (=crowd) as well as
professional analysts from banks, brokers and research companies.

Every stock prediction is visible to the other members of the platform. Besides the predictions of the Internet crowd, the website also collects the recommendations of leading banks such as HSBC, Goldman Sachs, Deutsche Bank or
Morgan Stanley. In addition, recommendations of brokers and research companies (e.g., Independent Research, Kepler) are also part of our dataset. Analysts of
these financial services companies will be referred to as "analysts" or "experts"
in the following analysis. Thus, a recommendation of an analyst always occurs in
the name of the respective company. While crowd members have to register on
the website and fill out the Internet form for publishing their recommendations,
the professional share recommendations are automatically integrated every time
a bank publishes a new share analysis. Overall, our dataset consists of 10,146
single stock predictions published between May 5, 2007 and August 15, 2011.
1,623 different crowd members made 8,331 recommendations whereas 40 different analysts (i.e. financial institutions) made 1,815 recommendations. These
numbers indicate that the crowd is much larger compared to the group of analysts. We only considered blue chip stocks from the DAX index to ensure that
stock predictions on the platform have no direct market impact and thus to avoid
endogeneity problems which may exist, for example, for penny stocks. The DAX
is the most important stock market index in Germany, containing the 30 largest
German companies. It is therefore comparable to the Dow Jones Industrial Average in the US.

In the same way as professional analysts operate in the real world, crowd
members can open and close their recommendations at any time during regular
trading days. Each recommendation is automatically closed after the maximum
duration of 180 days. Recommendations from professional analysts and the
crowd are presented in a similar manner (see Figure 3-1 for a screenshot). Beside

Figure 3-1: Screenshot of an Analyst's Recommendation

the name of the bank or crowd member, each recommendation consists of the rating (buy, sell or hold), current price, target price, start price, actual performance as well as target return. In addition, the website also shows information on the previous accuracy (ranking).

It is important to note that only the crowd members communicate with each other on the website. Members can write public comments on other recommendations, private messages to virtual friends or take part in forum discussions. Professional analysts are not an active part of the stock community rather their recommendations in the name of the bank are automatically integrated on the website as soon as these recommendations have been released to the public.

While other stock prediction communities also provide buy and sell ratings, this platform is unique in terms of the specification of price targets as well as the opportunity to close recommendations. So far, researchers had to choose a time horizon by their own (i.e., four weeks or two months), assuming that an open stock prediction is valid during the entire period. But it surely can make a difference if an investor opens a recommendation on one day and closes it three days after when his opinion has changed. The unique features of the community platform enables us to take potentially different durations into account and thus to precisely determine the performance and compare the results on a daily basis.

Table 3-1 provides descriptive statistics for all variables, which are used in the following analysis. We obtained stock market data from the website of the Frankfurt Stock Exchange (FWB)[2]. All prices and trading volumes which are used in the analysis refer to executed trades on the Frankfurt floortrading stock exchange[3].

---

2 www.boerse-frankfurt.com
3 Besides the Frankfurt floortrading exchange, there is also the electronic trading system XETRA. Both exchanges differ to the extent that on the floor, prices are determined by market makers while trades on XETRA are executed electronically. However, prices for DAX equities are almost identical since XETRA prices are the reference for all other regional exchanges in Germany, including the Frankfurt

Table 3-1:   Operationalization Summary

Variable	Unit	Min	Max	Mean	Std. Dev.
Daily Return	Daily return of recommended stocks	-0.06	0.20	0.0018	0.0134
Analysts	Dummy variable for the presence of professional analysts on the platform (0 = present; 1 = otherwise)	0	1	0.50	-
Ranking	Dummy variable for the improved ranking system on the platform (0 = improved ranking system is present when the recommendation is made; 1 = otherwise)	0	1	0.34	-
Age Diversity	Standard deviation of all crowd members' age	8.44	12.10	11.64	0.68
Gender Diversity	Gender diversity of all crowd members as measured by 1 - | share of male - share of female|	0	0.11	0.078	0.24
Momentum	Share price when recommendation is made divided by share price three months before	0.12	5.55	1.0099	0.26
Trading Volume	Average daily turnover (in €) of shares within the last three months before the stock pick	14,001	14,806,239	1,965,386	1,816,787
DAX Trend	Dummy variable for the DAX performance (=overall market trend) during the recommendation period (1 = bull market, i.e. level of the DAX increases during the recommendation period; 0 = otherwise)	0	1	-	-
Risk	Standard deviation of the daily returns of recommended shares within the observation period	0.015	0.039	0.025	0.007
Activity	Number of stock predictions divided by the period of membership on the platform prior to the recommendation	0	15	0.7916	1.2610
Accuracy	Share of accurate stock predictions prior to the recommendation	0	0.98	0.5024	0.2513

floortrading exchange. Trading volumes might differ between the exchanges but have a strong correlation. The platform uses floortrading data for determining the start and end prices of the recommendations, which is why we also use stock market data of the Frankfurt floortrading exchange.

In order to test our hypotheses, we use the daily return of recommended stocks as outcome variable of interest. Assume that a buy recommendation for BMW was opened on May 3 with a price target of 66€ for this particular share. Assume further that the recommendation was opened at 3 p.m. when the current share price of BMW was 60€. One month later on June 3, the recommendation was closed by the member. During this month the share price increased by 5 percent to 63€. Thus, this stock prediction for BMW would have achieved an overall return of 5 percent. In order to compute the daily return, we divide the overall return by the term of the recommendation (in days). Thus, in this case we divide 5 percent by 30 days and receive a daily return of 0.17 percent. We measure daily returns to make different time horizons comparable since it makes a difference if someone is able to achieve a 5 percent return within one month or six months. In case of a sell recommendation, an individual achieves a positive return if the share price decreases. Please note that the bid/ask spread is neglected when measuring the performance. Instead, we take the last price before the recommendation was opened or closed, i.e. the price at which the last trade between a buyer and a seller was executed at the Frankfurt Stock Exchange. Given the small bid/ask spreads for DAX equities and the relatively long recommendation periods, spreads should not play an important role in our case. This would be different if we focused on daytrading activities.

Two radical changes on the platform reduce the degree of independence among the members of the crowd. First, independence decreases after the publication of professional analysts' recommendations on the platform. These were added in October 2009 and allowed us to investigate whether the presence of professional analysts exerts an influence on the crowd's investment decision.

The second threat to independence is the introduction of a new ranking system in May 2010, which provides a more precise picture of the members' accuracy compared to the old system. Between 2007 and 2010, the rankings were only based on the hit rate (ratio between correct and wrong picks) and average performance of the recommendations. The revised ranking system provides several improvements so that the figures are more meaningful. Now, a complex algorithm calculates the rankings, ensuring a high degree of transparency and forecasting quality. Another new component is that the ranking calculation only considers shares fulfilling certain quality criteria. For example, the particular share must trade above .10 EUR and exceed a daily trading volume of 500,000 EUR, which ensures that so-called penny stocks are excluded from the calculation.

A further modification is that a member must reach a minimum number of five recommendations before receiving a ranking position. The algorithm then determines the members' skill level on a daily basis through carrying out buy or sell transactions in a virtual depot. The skill level is thereby calculated by the comparison between the performance of the virtual portfolio and the STOXX Europe 600, a broad European market index. In sum, performance indicators are more realistic now so that the improved ranking system provides a more precise picture of the members' ability to predict the stock market development. In addition to quality improvements, the platform provider made considerable efforts to introduce the ranking system to the community members (e.g., beta testers). Top users are more visible now since members with the highest prediction accuracy are marked with a "top user" symbol. We therefore suggest that more members will consider the other users' recommendations so that independence will decrease.

We further need information on the degree of diversity on the platform. H2 postulates that the performance of the crowd improves with greater diversity. Previous studies frequently operationalized diversity by means of demographic information, such as age and gender (see section 2). We follow this approach and operationalize diversity by the variance of age and gender distribution of the crowd members based on the self-reported personal profiles on the platform.

We define age diversity as the standard deviation of the age of all members, which is around 8 in May 2007. This value increases to 12 by the end of our observation period (see Appendix). We operationalize gender diversity by considering the deviation of the ratio between male and female members from the 50:50 gender ratio on the platform:

$$\text{Gender Diversity} = 1 - |\text{ Share of Male} - \text{Share of Female }| \qquad (3.1)$$

For instance, if there were a totally balanced gender distribution on the platform (50 percent men and 50 percent women), the deviation from the 50:50 ratio would be 0 so that gender diversity has the highest possible value of 1. The lowest value of 0 would occur if only men or women were registered. Thus, the higher the value for this variable, the more diverse is the platform with respect to gender.

At the end of 2007 there was only a gender diversity of 0.055, compared to 0.083 in August 2011. With a higher standard deviation of the crowd members' age (in years) and changing gender diversity, both diversity measures increase over time (see Appendix). Studies in financial economics have shown that stock returns depend on stock specific characteristics and overall market conditions. It is therefore necessary to control for a number of factors, which are not part of the WoC phenomenon. Momentum implies the performance within the last three months before the recommendation was opened. This figure simply shows whether a stock was a previous winner or loser. Infineon showed the highest and lowest three month momentum in our sample. The share price quintupled between March and June 2009 but lost 88 percent between September and December 2008.

The inclusion of momentum is necessary since the members' return might depend on market trends (upturn or downturn phase). In a similar study of a stock prediction community, Avery et al. (2009) also considered the stocks' momentum for distinguishing between bull and bear markets. Momentum strategies (buying past winners and selling past losers) are very common among investors. Many studies which investigate the stock picking ability of mutual fund managers refer to the momentum effect (e.g., Carhart 1997; Daniel et al. 1997). Among others, Jegadeesh and Titman (1993) documented that superior returns in the US stock market can be achieved by selecting shares based on their performance in the past 3 to 12 months. Rouwenhorst (1998) confirmed this return continuation for European countries. The literature provides different explanations for the relationship between past performance and future stock returns, for example data mining or behavioral patterns (Hong and Stein 1999).

Trading volume represents the average daily turnover of shares within the last three months before the recommendation was opened. There are companies, which are heavily traded especially considering the critical months of the financial crisis in 2008 and 2009 (i.e. Commerzbank, Deutsche Bank). On the other hand, companies of more defensive sectors (e.g., E.ON) or smaller companies (e.g., Infineon) experience a much smaller trading volume. The average daily turnover prior to the share recommendation is 2 million €.

DAXTrend is a dummy variable for the DAX performance (=overall market trend) during the recommendation period (1 = bull market, i.e. level of the DAX increases during the recommendation period; 0 = otherwise). Previous studies indicate that forecasting abilities of investors might depend on market trends, being more optimistic during bull markets and vice versa (see for example DeBondt 1993). We therefore control for the stock market climate on the macro level.

Risk shows each company share's risk as measured by the standard deviation of daily returns within the observation period. Returns of riskier stocks fluctuate more heavily and therefore have a higher standard deviation. We observe the lowest standard deviation for Deutsche Telekom (.015) while Infineon is the riskiest company in our sample (.039). Studies in the area of stock predictions typically take the risk of individual stocks into account (e.g., Hill and Ready-Campbell 2011).

Finally, we collected member specific characteristics on trading activity and accuracy. This information is calculated by the platform provider and only included when we compare the average performances between crowd members and analysts. Activity and accuracy measures refer to all stock picks of a member including smaller companies which are not part of the DAX index. In contrast to studying the influence of diversity and independence on the overall performance (macro view), the comparison between individual members of both groups is performed from a micro perspective so that member specific information is redundant for the latter research question. Activity is the number of stock predictions divided by the period of membership on the platform before the recommendation was opened. This measure indicates how active a community member has been before opening the recommendation. On average, members open .79 recommendations per day. Please note that this measure not only relates to DAX equities but to all recommendations a member has opened. The minimum number is obviously 0 since some members start with recommending DAX equities. Accuracy represents each member's forecast ability. This number is calculated by the platform provider and shows how many predictions have been correct in the past. We include both variables to isolate the characteristic of being a professional analyst when testing H1.

3.3.2   Data Analysis

Comparison of Forecast Accuracy between Professional Analysts and the Crowd

Since professional analysts and the Internet crowd might make their recommendations in different situations (self-selection bias), causal inferences are quite challenging and simple regression analyses are not suitable. Therefore, we use propensity score matching (Rosenbaum and Rubin 1983). Matching analyses are similar to regression models to the extent that both methods aim to draw causal inferences. We aim to compare the forecast accuracy between professional analysts and the crowd. Our independent variable is therefore the daily return of stock recommendations. One might compare both groups by conducting regression analysis or simply calculating the average performances. However, stock predictions probably not only differ with respect to group membership (i.e. recommendation of the crowd vs. analysts) but also with respect to other characteristics, such as market parameters. The propensity score matching approach aims to compare members of the treated population and non-treated members of the control group which resemble each other in all characteristics but the treatment. Thus, the reason for the group difference with respect to the variable of interest (daily return) can be exclusively identified by the treatment, which is the level of expertise in our case (professional analysts compared to crowd members). Every analyst recommendation that is published on the platform is part of the analyst group (="treated" population), while every crowd recommendation is part of the crowd group. Thus, the separation of both groups is achieved by identifying the person who opened the recommendation.

We first identify statistical twins with respect to characteristics, i.e. one recommendation of the crowd and a similar recommendation of the analysts. For this reason we first compute propensity scores, which represent the probability that a recommendation was made from a professional analyst given the following control variables: Trading volume, momentum, DAXTrend, risk, activity and accuracy (see section 3.3.1 for a description of all variables).

In a next step, we match the recommendations of analysts and the crowd. Only share recommendations that resemble each other in the above mentioned characteristics will be compared. Since it is almost impossible to find statistical twins, which have identical values for all characteristics, we calculate propensity scores in order to determine the similarity. Two share recommendations with similar propensity scores can be compared so that the results are unbiased and allow us to attenuate a potential self-selection bias. The reader is referred to Heckman et al. (1997) for a detailed description of the matching method.

Diversity and Independence

The structure of our data is similar to panel data in a way that every stock is repeatedly recommended over time. This allows us to define a panel variable representing each of the 30 DAX companies. We use a random effects model described by equation 2. The Hausman specification test (p>.05) denied the use of fixed effects and we therefore prefer to estimate the model under use of random effects which thus absorb company specific effects. Furthermore the Breusch-Pagan test indicated the presence of heteroskedasticity (p< .01) and we therefore estimate the model with robust standard errors.

We use the following equation in order to estimate the influence of diversity and independence on the daily return of recommended stocks:

$$
\begin{aligned}
\text{DailyReturn}_{i,t} = {} & ß_0 + ß_1 * \text{AgeDiversity}_t + ß_2 * \text{GenderDiversity}_t + ß_3 \\
& * \text{Ranking}_t + ß_4 * \text{Analysts}_t + ß_5 * \text{Momentum}_{i,t} + ß_6 \\
& * \text{TradingVolume}_{i,t} + ß_7 * \text{DAXTrend}_t + ß_8 * \text{Risk}_i + ß_9 \\
& * \text{Time}_t + \alpha_i + \varepsilon_{i,t}
\end{aligned}
$$

$$(3.2)$$

where DailyReturn_i is the daily return of the recommended stock i; AgeDiversity_i is the standard deviation of all registered crowd members' age when the recommendation of stock i is made; GenderDiversity_i as measured by 1 - | share of male – share of female | indicates how far the ratio between male and female members on the platform deviates from the 50/50 gender ratio when the recommendation of stock i is made; Ranking_i is a dummy variable for the improved ranking system on the platform (0 = improved ranking system is present when the recommendation of stock i is made; 1 = otherwise); Analysts_i

indicates whether analysts' recommendations are published on the platform or not (0 = present when the recommendation of stock i is made; 1 = otherwise); $Momentum_i$ indicates the performance of the stock i within the last three months; $TradingVolume_i$ represents the average turnover of the stock i within the last three months; $DAXTrend_i$ is a dummy variable for the DAX performance (=overall market trend) during the recommendation period (1 = bull market, i.e. level of the DAX increases during the recommendation period of stock i; 0 = otherwise); $Risk_i$ is the degree of a share i's risk as measured by the standard deviation of daily returns during our observation period; $Time_t$ is a variable which increases by 1 every day of the analysis in order to control for a time trend; α_i captures all specific characteristics of stock i which are constant over time and cannot be described by the control variables; $\varepsilon_{i,t}$ is the error term.

3.4 Results of Empirical Study

3.4.1 Comparison of Forecast Accuracy between Professional Analysts and the Crowd

Results from the probit regression (Table 3-2) reveal that the probability of being an analyst's recommendation decreases by trading volume, activity and risk. The probability increases by accuracy, DAXTrend and momentum.

Table 3-2:   Results from Probit Regression

	Coefficient	Std. Err.	p-value
Constant	-2.638	.107	<0.001
Momentum	.501	.066	<0.001
TradingVolume	-.000	.000	<0.001
DAXTrend	.627	.039	<0.001
Risk	-20.399	2.518	<0.001
Activity	-.145	.018	<0.001
Accuracy	2.756	.100	<0.001

Dependent variable: Probability of being an analyst's recommendation; Number of observations: 10,146; $R^2 = 0.210$

Table 3-3: Results of Propensity Score Matching

Depen-dent variable Daily return	Sample	Treated (Analysts)	Controls (Crowd)	Differ-ence	S.E.	T-stat
	Unmatched	.0009971	.0020340	-.001037	.000347	-2.99
	Average treatment effect on the treated (ATT)	.0009971	.0026229	-.001626	.000299	-5.43

We are now able to match stock predictions of analysts and the crowd based on similar propensity scores. Table 3-3 indicates that, on average, the daily return of an analyst is 0.0016 percentage points less than the return of a statistical twin of the crowd whose prediction is similar in terms of the control variables (p<0.01). Without the matching method, the difference between controls (=crowd) and treated (=analysts) would be underestimated (see "Unmatched" row in Table 3-3). This is why we have to match the recommendations.

Our results provide empirical evidence for the phenomenon that the crowd is able to outperform the experts with regard to the prediction of share price returns, supporting H1.

The difference between the average performance of the crowd and analysts is statistically significant (T = 5.43; p<0.01, two-tailed t-test) but economically small, even if we make a projection for the entire year. Assuming 365 days, Internet users achieve an annual return that is 0.59 percentage points higher than professional analysts. For the reader, this number should only serve as orientation since security returns are usually provided on annual basis. In our case, we only measure and compare the average return per day. The construction of trading portfolios which follow the members of the stock prediction community is out of the scope of this article. The reader is referred to Hill and Ready-Campbell (2011) who create sophisticated trading algorithms based on the crowd's stock picks and Gottschlich and Hinz (2013) who develop a decision support system based on crowd's recommendations.

However, we are confident that the crowd can also achieve reasonable returns after the consideration of transaction costs. The average duration of a crowd member's recommendation is 81 days, while analysts' recommendations are closed after 128 days. Thus, we need on average 8 trades per year (4 times open and close) for the crowd and 4 trades for analysts. Without transaction costs, (matched) crowd members realize a return of 0.0026 percent per day (or 0.95 percent per year), while analysts achieve 0.0010 percent per day (or 0.37 percent per year). Assuming transaction costs of 0.1 percent per trade, this would reduce the return of an average crowd member to 0.15 percent, while the overall return of analysts would even be negative (-0.03 percent).

Thus, the superiority of the crowd would remain even after taking transaction costs into account. However, from an economic point of view, these effects are quite small. We can therefore confirm previous results from Antweiler and Frank (2004) who studied the influence of stock message postings on returns and conclude that the result "does seem to be economically small but statistically robust" (pp. 1261). Our study not only indicates the existence of a collective intelligence within groups, but also finds evidence for the superiority of large diverse groups compared to a few experts.

Another interesting finding is that the crowd is also able to outperform the broader stock market. The DAX index lost 20 percent within the period of analysis and thus both, the focal community crowd and experts, outperformed the market substantially. At first glance this seems surprising given the fact that the size of the stock prediction community is relatively small compared to the entire stock market. The whole WoC approach is based on the assumption that large groups or markets perform better than smaller groups and individuals. According to the efficient market hypothesis, investors should not be able at all to gain superior returns (Fama 1970). However, our results confirm previous studies showing that a stock prediction community on the Internet is able to achieve excess returns against the stock market. In a study of Hill and Ready-Campbell (2011) the Internet crowd outperforms the S&P 500 by 12.3 percentage points in 2008. Earlier studies from Das and Chen (2007) and Antweiler and Frank (2004) focusing on Internet stock message boards also indicate that content from financial communities has predictive value to major stock indices.

One explanation for the superior performance of the Internet community compared to the overall stock market might be that the stock market is worse affected by negative influences that prohibit wise crowds. Numerous studies have shown herding among institutional investors (see section 3.2) and thus individual stock predictions of these professional analysts might not be independent enough from each other. As a result, speculative bubbles can evolve driving security prices far away from fundamental values since investors rely on common views of other investors instead of rationally evaluated market prices (Shiller 2002). Thus, the stock prediction community on the Internet might be more diverse and independent than the overall market which is characterized by extensive word-of-mouth communication (Hong et al. 2005).

With regard to the superior performance of the crowd compared to analysts, we identify one main explanation: the higher agility of the Internet users' stock recommendations. Agility figures of both groups are summarized in Table 3-4. We applied an independent-samples t-test for comparing mean values of the duration of recommendations. For the share of sell recommendations, we use the Fisher exact test due to the binary classification variable (Table 3-4).

First, crowd members are more active in opening and closing the recommendations. While the average duration of an analyst's prediction is 128 days, crowd members only leave their recommendations open for an average of 81 days. This is why the crowd is more able to take advantage of existing trends.

Another reason is the ratio between buy and sell recommendations. It is widely known that analysts prefer buy recommendations as every investor is able to buy shares, whereas in the case of a sell recommendation only the owners of the stock can respond to the recommendation. Furthermore the analyst is interested in maintaining a business relationship with the respective company and therefore wants to avoid negative evaluations, which are not popular for the management. The bank also might be interested in offering financial advisory to the rated company, i.e. capital increases or other investment banking-related services (Lakonishok and Maberly 1990).

There is much empirical evidence for the asymmetrical distribution of ratings (Barber et al. 2006; Dimson and Marsh 1986; Groth et al. 1979). For instance, Barber et al. (2006) found that in 1996 the number of sell recommendations from investment banks and brokerage firms was only 4 percent, declining to 2 percent in 2000 before increasing to 17 percent until 2003. Thus, even in times of stock market decline there is only a minority of sell recommendations.

For the purpose of our study, we do not directly compare the analysts' original ratings with the recommendations of the crowd since, in contrast to the analysts, crowd members cannot assign hold ratings on the platform. However, since every hold recommendation that we consider also has a price target, we can define a buy rating when the price target lies above the last price and a sell rating for the opposite case.

Our results show that only 24 percent of the analysts' predictions are sell recommendations in the sense that a lower future price is expected, while the crowd assigns the same in 27 percent of all cases. Given our observation period of four years with many downturns and upturns on the stock market, it is of little surprise that a more balanced distribution of buy and sell ratings seems to ensure a higher accuracy of stock predictions.

Table 3-4:   Comparison of Agility

Degree of agility	Group	N	Mean value	P > | t |
Share sell recommendations	Analysts	1815	24.46%	.018
	Crowd	8331	27.24%	
Duration of recommendation	Analysts	1815	127.90 days	.000
	Crowd	8331	80.89 days	

3.4.2   Diversity and Independence

Table 3-5 shows the results for the influence of diversity and independence on the performance of the Internet crowd. We estimated several models to ensure the robustness of our results. First, we only considered independence and then successively included diversity, market parameters and risk.

We observe a positive parameter for the crowd members' age as well as gender diversity. However, the results are not statistically significant except of gender diversity in model 2. If we control for market parameters and risk (full model), we can conclude that increasing diversity on the platform does not improve the daily return of the crowd, rejecting H2.

In contrast to diversity, we find evidence for the influence of independence on the performance of the crowd. The daily return is higher before analysts are present on the platform and before the new ranking system is introduced. The results are highly significant in all of our models. We therefore find support for H3.

The daily return of crowd members' stock predictions is 0.3 basis points (=0.003 percentage points) higher before recommendations of professional analysts are published on the website and 0.1 basis points higher before the revised ranking system is introduced. Crowd members achieve a higher return during bull markets (0.2 basis points) while increasing risk attitude is rewarded by 7.2 basis points. Surprisingly, momentum of individual stocks has a negative effect of 0.4 basis points, while trading volume exerts no significant influence on daily returns.

We observe a positive time trend, which is significant in all our models. This might be due to an increasing crowd size over time, which probably makes the entire crowd wiser. Another reason might be the increasing popularity of the stock prediction community, which ensures that more and more well-informed members join the platform, improving the quality of share recommendations. Finally, the structure of financial markets has changed between 2007 and 2011. For instance, there are more algorithmic traders in recent years compared to the pre-crisis period. However, we can only provide potential explanations but not fully explain the observed time trend.

We conducted several additional analyses in order to ensure the robustness of our results. First, we tested a daily-base Sharpe Ratio as dependent variable. The Sharpe Ratio was first introduced by Sharpe (1966) as a reward-to-volatility measure and is used by many authors under different names. The original Sharpe Ratio is calculated as follows:

$$\text{Sharpe Ratio} = \frac{(R_a - R_b)}{\sigma} \qquad (3.3)$$

where R_a is the return of an asset; R_b is the return of a benchmark investment (typically a riskless investment as measured by the risk-free interest rate); $(R_a - R_b)$ is the excess return and σ is the standard deviation of the excess return.

For our purpose we use a daily-base Sharpe Ratio to apply the measure for daily returns:

$$\text{Daily} - \text{base Sharpe Ratio} = \frac{R_r}{\sigma} \qquad (3.4)$$

where R_r is the daily return of the recommendation and σ is the standard deviation of daily returns during our observation period. The higher the daily-base Sharpe Ratio, the higher the return that was achieved per unit of risk.

Again, we estimate the models under use of random effects (see section 3.3.2). Please note that according to equation 3.4, risk is now included in the dependent variable.

$$\begin{aligned}
\text{Daily} &- \text{base Sharpe Ratio}_{i,t} \\
&= \beta_0 + \beta_1 * \text{AgeDiversity}_t + \beta_2 * \text{GenderDiversity}_t + \beta_3 \\
&\quad * \text{Ranking}_t + \beta_4 * \text{Analysts}_t + \beta_5 * \text{Momentum}_{i,t} + \beta_6 \\
&\quad * \text{TradingVolume}_{i,t} + \beta_7 * \text{DAXTrend}_t + \beta_8 * \text{Time}_t + \alpha_i \\
&\quad + \varepsilon_{i,t}
\end{aligned}$$

$$(3.5)$$

Table 3-6 shows the results for the daily-base Sharpe Ratio. Overall, our results with respect to diversity and independence do not substantially change. The positive and highly significant effect of independence persists while diversity does not exert a significant impact on the risk-adjusted daily returns.

To check the robustness of our results, we also modified the calculation basis for momentum, trading volume and risk. The period for momentum and trading volume was changed to one month and 12 months respectively. We observe similar results for daily returns as well as daily-base Sharpe Ratio compared to our original model (see Appendix, Tables 3-7 to 3-10).

In our full model (Table 3-5), risk is calculated by the standard deviation of daily share returns during our entire observation period of roughly four years. Thus, the individual risk for each share does not change over time. However, we modified the calculation basis for risk similar to the market parameters. That is, risk is measured by the standard deviation of daily returns within one month (and three months respectively) before the recommendation was opened (see Appendix, Table 3-11 to 3-14). In addition, we derived risk from the standard deviation

Table 3-5: Results from Regression Analysis (Dependent Variable: Daily return)

	Model 1: Independence	Model 2: + Diversity	Model 3: + Market parameters	Model 4 (full Model): + Risk
Analysts (0/1)	.003***	.003***	.003***	.003***
Ranking (0/1)	.002***	.001***	.001***	.001***
AgeDiversity		.000	.000	.000
GenderDiversity		.030**	.012	.011
Momentum			-.004***	-.004***
TradingVolume			.000	-.000
DAXTrend			.002***	.002***
Risk				.072***
Time Control	.000***	.000***	.000***	.000***
Observations	8,331	8,331	8,331	8,331
R^2	0.004	0.005	0.013	0.014

** Significant at the 5% level; *** Significant at the 1% level;
All models are estimated using random effects

Table 3-6: Results from Regression Analysis (Dependent Variable: Daily-base Sharpe Ratio)

	Model 1: Independence	Model 2: + Diversity	Model 3 (full model): + Market parameters
Analysts (0/1)	.143***	.145***	.148***
Ranking (0/1)	.064***	.039***	.054***
AgeDiversity		.001	.000
Gender Diversity		.937*	.262
Momentum			-.164***
Trading Volume			-.000
DAXTrend			.067***
Time Control	.000***	.000***	.000***
Observations	8,331	8,331	8,331
R^2	0.005	0.006	0.015

* Significant at the 1% level;** Significant at the 5% level;
*** Significant at the 1% level;
All models are estimated using random effects

of daily share returns between the start and the end of the recommendation (see Appendix, Tables 3-15 and 3-16).

All additional analyses confirm our original results, observing a positive effect of independence on daily returns as well as daily-base Sharpe Ratio.

It should be noted that on first sight, R^2 values presented in Table 3-5 and 3-6 indicate a rather weak explanatory power of our model. However, researchers studying the drivers of stock returns typically report very low R^2 values (e.g., Antweiler and Frank 2004; Avery et al. 2009; Das and Sisk 2003; Malmendier and Shanthikumar 2007). Similar to Das and Chen (2007) who explain the poor overall fit of their model with the fact that the "regression lacks several other variables that explain stock levels", we also argue that share prices depend on many factors which are not part of our regression analysis. For instance, we do not control for company or economic news, which have been found to exert a strong influence on share prices (e.g., Mitchell and Mulherin 1994; Niederhoffer 1971; Tetlock 2007). Recently, Goh et al. (2013) estimated the influence of user-generated content on consumer behavior under use of random effects. The authors justify low R^2 values in arguing that their research "does not involve forecasting, thus R^2 model fit may matter less." We therefore believe that our study fits well with the existing stream of research, contributing to a better understanding of the WoC on the Internet.

3.5 Discussion

3.5.1 Implications

Our results have strong implications for the financial service industry as well as companies from other industries. From a practical point of view, the financial service industry can take the opinion of the crowd into consideration for their investments. Given today's inflexible system of share analysis, private investors are on average better served by trusting the recommendations of an online prediction community instead of following the advice of their banks' analysts. One possible way to take advantage of the user-generated content is to create a portfolio which is based on the crowd's stock recommendations. Banks might issue investment funds that reproduce the buy or sell recommendations of the leading crowd members and thus develop real-time trading strategies. According to our study, it can be expected that the performance will be superior to the broader market as well as many conventional investment funds that are based on the analysts' recommendations. However, those funds might be vulnerable to manipulation by crowd members, especially if the crowd consists of a very large and unknown population. On the other hand, if the strategy only reproduces the stock picking behavior of too few individuals, the crowd's wisdom might disappear. Any construction of investment products must therefore ensure that a) the crowd size is appropriate and b) crowd members cannot manipulate the strategy.

Our study contributes to the debate about the WoC in such a way that independence seems to be an important condition on the Internet. The performance of the crowd positively relates to a higher degree of independence. Companies which employ crowdsourcing and open innovation concepts should thus ensure that decisions are made independent from each other. In light of our results, independence is especially important for the area of finance. We know from the offline world that converging investment behavior can destabilize security prices, resulting in lower returns for investors in the long run (see discussion on herding in section 3.2). Our study provides evidence for similar effects on the Internet. Thus, crowd members should primarily rely on private information and follow their own beliefs instead of trusting other market participants.

3.5.2   Summary and Outlook

This field study revealed that the WoC phenomenon that has been widely discussed by researchers and popular science authors can be observed on the Internet, but it must be approached on a differentiated basis. User-generated content undoubtedly contains valuable information that might increase market efficiency and overall welfare. For instance, the crowd is able to make better stock market predictions than professional analysts from banks, brokers and research companies. On an annual basis, the crowd realizes a 0.59 percent higher return than analysts.

While our field study confirms previous results with regard to the accuracy of Internet applications (Forsythe et al. 1999; Ginsberg et al. 2009; Spann and Skiera 2003), we only partly find support for the postulated theoretical conditions that have been found to be necessary for a wise crowd in the offline world. Knowledge, motivation, diversity and independence of the community members on our observed platform seem to be significant enough to create crowd wisdom although we did not measure these conditions in absolute terms.

We particularly conclude that the performance of the crowd improves with a higher degree of independence. We therefore find support for the importance of independence in the online world and confirm previous results from Lorenz et al. (2011) who experimentally showed that little social influence is enough to eliminate the WoC. As expected, the daily return of the crowd decreases after the introduction of recommendations made by professional analysts. The revised ranking system, which makes top performer more visible and shows a much more precise picture of the members' ability, also exerts a significant influence on the crowd's performance. In both cases, members of the community seem to increasingly rely on the opinions of so-called experts because of the assumption that highly paid analysts and the crowd's top performer have more knowledge or stock picking skills.

Empirical evidence indeed suggests that people attach great importance to the opinion of experts. For instance, courts place reliance on the psychiatric predictions with regard to patients' potential dangerousness, although many previous studies show that psychiatrics are not able to forecast the patients' behavior (Cocozza and Steadman 1978).

With the exception of model 2, we find no evidence for the influence of diversity. The missing effect of gender diversity might be caused by the very small fraction of females on the platform (<5 percent). Age diversity seems to play only a minor role on the Internet in contrast to the offline world (Bantel and Jackson 1989). The rejection of hypothesis 2 has to be put into perspective to the extent that age and gender diversity only slightly increase over time, remaining relatively constant for most parts of our observation period. We only used age and gender to operationalize diversity in a specific financial markets environment. Future research might consider other diversity aspects (e.g., knowledge, education, etc.) and especially verify if our results hold for areas outside the financial industry.

Our analysis is restricted to the exclusive consideration of blue chip stocks from the DAX index. With the focus on large companies, we are able to avoid endogeneity problems since it is not to be expected that single recommendations or comments of Internet users will directly influence the price of these stocks. However, the consideration of small and mid-sized companies would be an interesting area for future research since such organizations allow investors to better take advantage of private information. Only a few analysts cover smaller stocks and therefore company-related information is typically processed more slowly by the media and other investors.

We examined the impact of a changing degree of independence (ranking system and analyst recommendations) and diversity (age and gender diversity) on prediction accuracy. However, these measures are relative and we cannot exactly determine an absolute level of diversity and independence. Future research might try to address this problem in experimental settings. With the help of experiments, one could also eliminate the limitation of restricted access to information. Field studies typically have the problem that the operationalization is partly driven by the available dataset. While there are various other variables that have been used in the past in order to measure diversity (see section 3.2), we have only access to the age and gender that the members are providing in their personal profiles. The data does not allow for making conclusions about knowledge, education or other distinguishing factors. The same limitation holds for the independence variables. Thus, the operationalization would certainly benefit from an experimental setting in future research projects.

With regard to the composition of the crowd, we are not able to draw conclusions about the share of expertise. There may be professional analysts that register anonymously on the platform and open recommendations in their spare time. We describe the crowd as dedicated amateurs, i.e. sidelined investors who may be more or less professional. However, we have to be careful when interpreting the superior performance of the crowd compared to analysts since the result might be diluted by a certain fraction of analysts within the crowd. Transaction costs might also reduce the performance of both groups. In our study, we only compare individual stock predictions without developing trading portfolios. An interesting area for future research would be to create algorithms which buy or sell shares according to the stock picks of the crowd and analysts. Since members can write messages and comment on other recommendations, interaction processes would also be an avenue for further research.

Overall, we have to be careful when interpreting our results. The stock predictions might not represent the true opinion of the crowd members or analysts. Especially the crowd recommendations could be uninformative babbling ("cheap talk") since sharing private information might reduce profits from stock returns (Bennouri et al. 2011). The informational value of the recommendations could also be linked to price manipulations. Although we are aware of this problem and therefore excluded penny stocks from our analysis, future research might additionally look at the real stock portfolios of investors. This would allow researchers to clearly identify attempts where members try to push prices after buying stocks in the real world.

The recommendations of professional analysts might be influenced by business interests. Analysts publish their recommendations in the name of the bank which has strong incentives for assigning buy ratings due to business relationships with the respective companies. Another reason for the preference of buy instead of hold (or sell) ratings is the trading volume since banks benefit from high trading volumes. Optimistic share recommendations address much more clients than sell recommendations since only a small fraction of investors already has the shares. Thus, inferring a superior performance of the crowd might be a result of strategic constraints. However, the better performance of the crowd compared to analysts might have implications for private investors behavior since many investors still rely on analyst recommendations (Kelly et al. 2012; Malmendier and Shanthikumar 2007).

In sum, this study provided evidence that the WoC phenomenon exists on the Internet, but not all findings from conceptual work and experiments with regard to the necessary conditions can blindly be transferred. The WoC phenomenon turns out to be very complex, which underlines the need for more research in this area.

3.6 Appendix

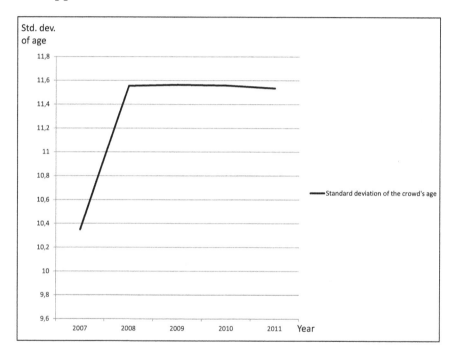

Figure 3-2:   Development of Standard Deviation of Age over Time

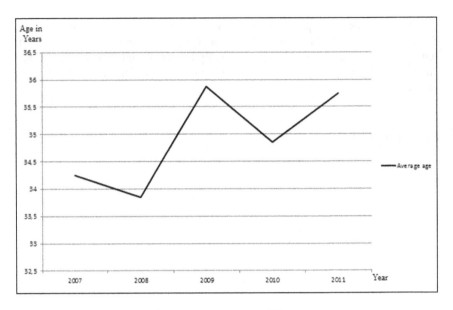

Figure 3-3:   Development of the Average Age over Time

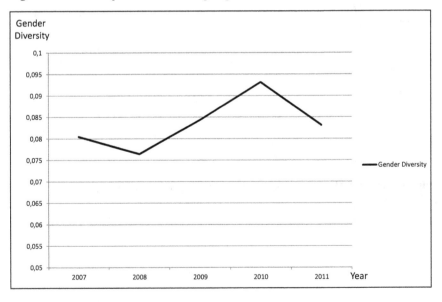

Figure 3-4:   Development of Gender Diversity over Time

Original model estimated with 1 and 12 month momentum / trading volume

Table 3-7: Results from Regression Analysis (Dependent Variable: Daily return)

	Independence	Diversity	Market parameters	Risk
Analysts (0/1)	.003***	.003***	.003***	.003***
Ranking (0/1)	.002***	.001***	.001***	.001***
AgeDiversity		.000	.001	.001
GenderDiversity		.030**	.004	.003
Momentum 1month			-.006***	-.006***
TradingVolume 1month			.000	.000
DAXTrend			.002***	.002***
Risk				.055***
Time Control	.000***	.000***	.000***	.000***
Observations	8,331	8,331	8,331	8,331
R^2	0.004	0.005	0.013	0.014

** Significant at the 5% level; *** Significant at the 1% level; All models are estimated using random effects

Table 3-8: Results from Regression Analysis (Dependent Variable: Daily return)

	Independence	Diversity	Market parameters	Risk
Analysts (0/1)	.003***	.003***	.003***	.002***
Ranking (0/1)	.002***	.001***	.001***	.001***
AgeDiversity		.000	.000	.000
GenderDiversity		.030**	.022*	.020
Momentum 12months			-.000	-.001*
TradingVolume 12months			-.000	-.000*
DAXTrend			.001***	.001***
Risk				.075***
Time Control	.000***	.000***	.000**	.000*
Observations	8,331	8,331	8,331	8,331
R^2	0.004	0.005	0.007	0.014

* Significant at the 10% level; ** Significant at the 5% level; *** Significant at the 1% level; All models are estimated using random effects

Table 3-9:   Results from Regression Analysis (Dependent Variable: Daily-base Sharpe Ratio)

	Independence	Diversity	Market parameters
Analysts (0/1)	.143***	.147***	.149 ***
Ranking (0/1)	.064***	.041***	.052***
AgeDiversity		.001	.019
GenderDiversity		.937*	-.099
Momentum 1 month			-.256 ***
TradingVolume 1 month			.000
DAXTrend			.071 ***
Time Control	.000***	.000***	.000***
Observations	8,331	8,331	8,331
R^2	0.005	0.006	0.015

* Significant at the 10% level; ** Significant at the 5% level; *** Significant at the 1% level; All models are estimated using random effects

Table 3-10:   Results from Regression Analysis (Dependent Variable: Daily-base Sharpe Ratio)

	Independence	Diversity	Market parameters
Analysts (0/1)	.143***	.147***	.116 ***
Ranking (0/1)	.064***	.040***	.040 ***
AgeDiversity		.001	.008
GenderDiversity		.937*	.621
Momentum 12 months			-.022
TradingVolume 12 months			-.000
DAXTrend			.061***
Time Control	.000***	.000***	.000**
Observations	8,331	8,331	8,331
R^2	0.005	0.006	0.009

* Significant at the 10% level; ** Significant at the 5% level; *** Significant at the 1% level; All models are estimated using random effects

Results for risk as measured by 1 and 3 month standard deviation of daily share returns

Table 3-11: Results from Regression Analysis (Dependent Variable: Daily return)

	Independence	Diversity	Market parameters	Risk
Analysts (0/1)	.003***	.003***	.003***	.002***
Ranking (0/1)	.002***	.001***	.001***	.001***
AgeDiversity		.000	.001	.000
GenderDiversity		.030**	.004	.001
Momentum 1month			-.006***	-.006***
TradingVolume 1month			.000	.000
DAXTrend			.002***	.002***
Risk 1 month				.075***
Time Control	.000***	.000***	.000***	.000***
Observations	8,331	8,331	8,331	8,331
R²	0.004	0.005	0.013	0.017

** Significant at the 5% level; *** Significant at the 1% level;
All models are estimated using random effects

Table 3-12: Results from Regression Analysis (Dependent Variable: Daily return)

	Independence	Diversity	Market parameters	Risk
Analysts (0/1)	.003***	.003***	.003***	.002**
Ranking (0/1)	.002***	.001***	.001***	.001**
AgeDiversity		.000	.000	-.000
GenderDiversity		.030**	.012	.014
Momentum 3months			-.004***	-.004***
TradingVolume 3months			.000	-.000
DAXTrend			.002***	.002***
Risk 3months				.079***
Time Control	.000***	.000***	.000***	.000**
Observations	8,331	8,331	8,331	8,331
R²	0.004	0.005	0.013	0.016

** Significant at the 5% level; *** Significant at the 1% level;
All models are estimated using random effects

Table 3-13:   Results from Regression Analysis (Dependent Variable: Daily-base Sharpe Ratio)

	Independence	Diversity	Market parameters
Analysts (0/1)	.062**	.068**	.070**
Ranking (0/1)	.065***	.049***	.058***
AgeDiversity		.001	.000
GenderDiversity		.850*	.009
Momentum 1 month			-.204***
TradingVolume 1 month			.000
DAXTrend			.058***
Time Control	.000***	.000***	.000***
Observations	8,331	8,331	8,331
R^2	0.004	0.004	0.010

* Significant at the 10% level; ** Significant at the 5% level; *** Significant at the 1% level; All models are estimated using random effects;Note: 1 month standard deviation used for calculation of daily-base Sharpe Ratio

Table 3-14:   Results from Regression Analysis (Dependent Variable: Daily-base Sharpe Ratio)

	Independence	Diversity	Market parameters
Analysts (0/1)	.064**	.074**	.077**
Ranking (0/1)	.053***	.036**	.049***
AgeDiversity		..000	-.001
GenderDiversity		1.041**	.563
Momentum 3 months			-.134 ***
TradingVolume 3 months			-.000
DAXTrend			.048**
Time Control	.000***	.000***	.000**
Observations	8,331	8,331	8,331
R^2	0.003	0.003	0.009

** Significant at the 5% level; ***Significant at the 1% level; All models are estimated using random effects; Note: 3 month standard deviation used for calculation of daily-base Sharpe Ratio

Results for risk as during recommendation period (start until the end of recommendation)

Table 3-15: Results from Regression Analysis (Dependent Variable: Daily return)

	Independence	Diversity	Market parameters	Risk
Analysts (0/1)	.003***	.003***	.003***	.002***
Ranking (0/1)	.002***	.001***	.001***	.001***
AgeDiversity		.000	.001	.000
GenderDiversity		.030**	.004	.006
Momentum 1month			-.006***	-.005 ***
TradingVolume 1month			.000	.000
DAXTrend			.002***	.002***
Risk				.063***
Time Control	.000***	.000***	.000***	.000***
Observations	8,331	8,331	8,331	8,331
R²	0.004	0.005	0.013	0.016

** Significant at the 5% level; *** Significant at the 1% level; All models are estimated using random effects; Note: Risk measured by standard deviation of daily share returns during recommendation period

Table 3-16: Results from Regression Analysis (Dependent Variable: Daily return)

	Independence	Diversity	Market parameters	Risk
Analysts (0/1)	.003***	.003***	.003***	.002***
Ranking (0/1)	.002***	.001***	.001***	.001***
AgeDiversity		.000	.000	.000
GenderDiversity		.030**	.012	.013
Momentum 3months			-.004***	-.003***
TradingVolume 3months			.000	-.000
DAXTrend			.002***	.002***
Risk				.064***
Time Control	.000***	.000***	.000***	.000***
Observations	8,331	8,331	8,331	8,331
R²	0.004	0.005	0.013	0.016

** Significant at the 5% level; *** Significant at the 1% level; all models are estimated using random effects; Note: Risk measured by standard deviation of daily share returns during recommendation period

4 Using Twitter to Predict the Stock Market: Where is the Mood Effect?[1]

Abstract

Behavioral finance researchers have shown that the stock market can be driven by emotions of market participants. In a number of recent studies mood levels have been extracted from Social Media applications in order to predict stock returns. We try to replicate these findings by measuring the mood states on Twitter. Our sample consists of roughly 100 million tweets that have been published in Germany between January, 2011 and November, 2013. In our first analysis we do not find a significant relationship between aggregate Twitter mood states and the stock market. However, in further analyses we also consider mood contagion by integrating the number of Twitter followers into the analysis. Our results show that it is necessary to consider the spread of mood states among Internet users. Based on our results in the training period, we created a trading strategy for the German stock market. Our portfolio increases by up to 36 percent within a six-month period after the consideration of transaction costs.

4.1 Introduction

Social Media has become a buzz word in public discussions, gaining more and more attraction from both academia and industry in the last years. In this article, we follow Kaplan and Haenlein (2010) who define Social media as a "group of Internet-based applications that build on the ideological and technological foundations of Web 2.0, and that allow the creation and exchange of User Generated Content" (pp. 61). This term includes all the well-known websites where people share their thoughts, pictures or videos with the Internet community (e.g., Facebook, Twitter, Google+, Youtube).

The number of people engaging with Social Media largely increased in recent years. According to Trendstream's Q4 2012 Global Web Index (Kosner 2013), 693 million people are active users on Facebook, followed by Google+

---

1 Nofer, Michael / Hinz, Oliver (2014). Using Twitter to Predict the Stock Market: Where is the Mood Effect? Business & Information Systems Engineering, forthcoming.

(343 million), Youtube and Twitter (both ~ 280 million). eMarketer (2012) expects the number of people visiting a social network site once per month to increase up to 1.83 billion by 2014. These numbers indicate that virtually any Internet user is engaging with Social Media today.

The value of user-generated content in terms of business forecasts has been shown in the literature. For instance, online consumer reviews can be used to predict movie success (e.g., Chintagunta et al. 2010; Dellarocas et al. 2007), video game sales (Zhu and Zhang 2010), music sales (Heimbach and Hinz 2012), or book sales (Chevalier and Mayzlin 2006).

There already exists some research investigating the influence of user-generated content on stock returns. Generally, one can distinguish between sentiment detection with respect to specific objects of interest and the analysis of mood levels, i.e. the strength of positive or negative mood states. Former methods for example focus on measuring the company sentiment by analyzing consumer reviews (e.g., Tirunillai and Tellis 2012) or content from stock message boards (Antweiler and Frank 2004). Twitter was also used to extract sentiment with respect to commodity markets and currency rates (Rao and Srivastava 2012).

While these approaches aim to determine the degree of positivity or negativity towards a firm or product, this article will primarily deal with the second approach, the analysis of mood levels. We will use Twitter to determine mood states on a general level. Behavioral finance and neurofinance researchers try to explain the link between investors' emotions and their trading behavior (e.g., Tseng 2006). For instance, individuals tend to be loss-averse, which means that they value losses more heavily than gains (Tversky and Kahneman 1991).

While early research was typically done in experimental settings, Social Media applications can now help to reveal the social mood (Nofsinger 2005). Individuals in good mood are more willing to invest in risky assets, such as stocks (Johnson and Tversky 1983). Thus, stock returns depend on the investors' risk appetite which in turn depends on their mood states.

The impact of feelings and emotions on the stock market was measured by using Twitter (e.g., Bollen et al. 2010), Facebook (Karabulut 2011) or LiveJournal (Gilbert and Karahalios 2010). The prediction of share returns based on mood states can be seen as market anomaly contradicting the efficient market hypothesis (e.g., Kamstra et al. 2000).

However, virtually no study considered social interactions of Internet users when showing the relationship between mood levels and stock returns. We therefore aim to extend previous research by including the number of Twitter followers into the analysis. The importance of every tweet depends on the number of users recognizing the original message. There is wide evidence that lead-users exert a large influence on other members of the community. Studies have also shown mood contagion, i.e. the transfer of emotions from leaders to followers (Bono and Ilies 2006; Sy et al. 2005) or between persons in general (Neumann and Strack 2000). A number of recent studies found evidence for emotional contagion on the Internet (e.g., Coviello et al. 2014; Guillory et al. 2011; Kramer et al. 2014). According to these findings, mood states can spread among Internet users through text-based communication.

First, we study the influence of changing social mood levels on share returns without considering the community structure. This enables us to answer the question if mood effects, which have been found by other researchers before, still exist in today's financial markets. There might be diminishing effects in recent years due to potential data mining strategies of investors. Afterwards, we include the importance of each tweet as measured by the number of followers into the analysis. It will become clear whether the predictive ability of mood states might be improved by considering social interactions of Internet users. After investigating the relationship between the social mood and the stock market in the training period, we apply a trading strategy to a different time period. Results of our virtual portfolio will show whether investors can actually profit from mood states in monetary terms.

In the next section we motivate our hypotheses and present previous research which investigated the influence of emotions on stock returns. We then describe the empirical study including the calculation of the Social Mood Indices (SMI and WSMI), our data set, method and results. On the basis of our results in the training period, we create a trading strategy for the German stock market. The paper concludes with a brief summary as well as implications for researchers and practitioners.

4.2 Previous Research

4.2.1   Behavioral Finance

Since the early 1990s behavioral finance researchers have continuously shown that the stock market is driven by investors' psychology. Investors are human beings who are prone to errors or at least emotion-based decisions[2]. Market anomalies were observed which contradict the efficient market hypothesis (Fama 1970) according to which the prediction of share prices should not be possible since market prices reflect every available piece of information.

---

2   Daniel et al. (2002) provide an extensive literature review showing that investors' psychology exerts an influence on security prices.

For instance, calendar anomalies refer to seasonal movements of stock market returns. The January effect means that returns are on average higher in January compared to other months of the year (e.g., Thaler 1987). One reason for this anomaly might be tax-loss selling. Investors aim to avoid taxes by selling shares, which have performed badly over the year. Then, at the beginning of the year, share prices recover from such selling pressure (Brown et al. 1983). Researchers also identified the Monday effect (also known as day of the week effect or weekend effect), implying that returns on Monday are relatively little to those on Friday before (e.g., Jaffe et al. 1989).

Anomalies can also be technical-related. The momentum effect implies that past winners (losers) continue to perform well (bad). This has been observed for single stocks (Jegadeesh and Titman 1993) as well as for indices (Chan et al. 2000). Investors also use the past performance of mutual funds as an indicator for future returns although persistence cannot be expected according to the efficient market hypothesis (Grinblatt et al. 1995).

Researchers provide different explanations for these market inefficiencies. Reasons for technical- and calendar-related anomalies are out of the scope of this paper. Instead, we focus on anomalies which are driven by feelings and emotions.

Behavioral finance researchers refer to two groups of investors which are important for the pricing information. First, rational arbitrageurs are well-informed investors who are not prone to sentiment. This group of investors is also known as "smart money" in the literature (De Long et al. 1990). On the other hand, noise traders irrationally rely on sentiment and other non-fundamental information which is unimportant in the eyes of rational traders (Black 1986). These noise traders follow trends and oftentimes over- or underreact to news.

The proponents of the efficient market hypothesis argue that rational arbitrageurs trade against noise traders, driving prices immediately back to fundamental values after exogenous shocks. Noise traders can therefore influence prices only for a very short time before rational traders take positions against them until the market equilibrium is reached (Fama 1965).

However, behavioral finance researchers have shown that the power of rational arbitrageurs in trading against noise traders is limited. De Long et al. (1990) refer to positive-feedback strategies: more and more noise traders might follow other noise traders when buying or selling stocks. In this case, noise traders buy (sell) in case of rising (falling) prices. Thus, rational speculators can anticipate the behavior of tomorrow's noise traders and also buy the stocks today, driving prices even higher.

There are a number of other factors, which limit the ability of rational investors to trade against the uninformed individual investors. For instance, smart money might have short selling constraints and other trading risks (Shiller 2003).

Since rational investors are mostly risk averse, the fundamental risk (e.g., vari-ance of share values) can also prevent arbitrageurs from trading for a certain period of time. The overall conclusion from this line of research is that sentiment can influ-ence share prices in case of limited arbitrage. Different sentiment measures have been proposed in order to forecast share returns, such as investor and consumer surveys (Brown and Cliff 2005; Lemmon and Portniaguina 2006; Qiu and Welch 2006), trading volume (Baker and Stein 2004) or market volatility (Whaley 2000). In this article, we focus on mood levels which have also been used as proxy for the investors' sentiment (Baker and Wurgler 2007).

4.2.2 Influence of Mood on Share Returns

According to neuropsychologists, mood is influenced by different factors. While dopamine was found to mediate the cognitive effects of positive mood, serotonin may be responsible for negative mood (Mitchell and Philipps 2007). During the day, not only events and stress levels influence people's mood states (van Eck et al. 1998) but also social interactions with other people (Vittengl and Holt 1998).

The literature reports many examples for mood-related anomalies. Saunders (1993) studied the period between 1927 and 1989 and found that stock returns at New York Stock Exchange are lower on cloudy days than on sunny days. The weather effect was confirmed by Hirshleifer and Shumway (2003) who show that sunshine is positively correlated with returns in 26 countries between 1982 and 1997. Both studies argue that sunshine creates good mood, which in turn affects investment behavior.

Sport events can also influence people's mood levels (Wann et al. 1994). Following this intuition, Edmans et al. (2007) studied the effect of international soccer game results on stock returns. The authors observe that domestic stock markets negatively react to losses of national soccer teams in international com-petitions (i.e. World Cup, Asia Cup, etc.). For instance, elimination from the World Cup leads to abnormal stock returns of 49 basis points on the next trading day. This loss effect holds for other sports, such as cricket or basketball. Chang et al. (2012) show that NFL game outcomes influence returns of companies, which are locally headquartered, confirming results on the national level.

Besides sport events or weather conditions, sleeping habits are another area of interest for studying the influence of emotions on asset prices. Kamstra et al. (2000) refer to the so called "daylight saving anomaly", which means that Mon-days after daylight-savings-weekends have lower stock returns than regular Mondays over the year. The reason for poorer returns lies in the fact that indi-viduals tend to shy away from risky assets due to increased anxiety which is caused by losses or gains of sleep.

Investors' mood might also be influenced by the level of air pollution. According to Levy and Yagil (2011), regions with a higher degree of air pollution (as measured by the Air Liquidity Index) show smaller returns compared to ecologically cleaner areas. Finally, Kamstra et al. (2003) investigated the role of depressions on investment behavior. Many individuals (and thus investors) suffer from seasonal affective disorder (SAD) during autumn and winter months when sunshine hours are scarce. Consequently longer nights lead to significantly lower returns for a number of stock markets in the world. The SAD effect was observed to be more pronounced in countries with a long distance to the equator (e.g., Sweden).

Thus, single events (e.g., sport results, daylight saving anomaly) or continuous effects (e.g., weather effect, daylight saving anomaly, air pollution) influence people's emotions. These mood-related anomalies can be explained by the misattribution bias according to which people make risky decisions depending on mood states (Johnson and Tversky 1983). Individuals in good mood are more optimistic with respect to uncertain future events. A person's emotional well-being is therefore important for subjective probability evaluations (Wright and Bower 1992).

The relationship between positive and negative mood states and the risk-taking tendency can be explained by the Affect Infusion Model (AIM) which postulates that people in positive mood rely on positive cues to make decisions (Forgas 1995). Because of the mood priming effect, people in positive moods associate risks positively in contrast to people in negative mood. Thus, the risk-taking tendency is higher for people in positive moods since they use heuristics and perceive the consequences of risky situations as more positive. People in negative moods are more prone to see the danger and are thus more careful in the decision process. Therefore they shy risks due to the negative associations with the risky decision (Schwarz 1990).

The AIM was confirmed by a number of laboratory experiments. For instance, Yuen and Lee (2003) induced subjects to positive and negative mood by showing corresponding movie clips. Results reveal that people in a bad mood show a more conservative risk-taking behavior compared to people in neutral or positive mood. Using a similar method, Chou et al. (2007) also report a higher risk-taking tendency for people in good mood compared to those in bad mood.

Depressive mood states have also been widely studied in the literature, especially by linking depression to levels of "sensation seeking", which is another measure for risk-taking tendency (e.g., Zuckerman 1984). It has been shown that depressive subjects have reduced sensation seeking compared to normal people (Carton et al. 1992). Bell et al. (2001) found that differences in risk behaviors can be explained by the levels of sensation seeking. Wong and Carducci (1991)

Figure 4-1:   Theoretical Framework

show that high sensation seekers have a greater risk-taking tendency in financial decisions than people with lower scores of sensation seeking. Furthermore, Eisenberg et al. (1998) show that depression correlates with risk aversion.

We argue that mood fluctuations influence the risk attitude, which in turn exerts an influence on the willingness to invest in risky assets, such as stocks (Figure 4-1). This relationship was shown in the above cited studies of behavioral finance. Stock returns are therefore expected to be influenced by mood states of market participants.

4.2.3   Predictive Value of Social Media

While earlier research used exogenous factors as variables of interest (e.g., weather, sport results), Social Media applications allow researchers now to precisely measure mood fluctuations by analyzing people's statements about their emotional well-being.

In a seminal work, Bollen et al. (2010) have shown that mood levels extracted from public tweets have predictive value to the Dow Jones Industrial Average (DJIA). At a time when the overall mood is calm (or to some extent happy), the authors find statistically significant evidence for an associated reaction of the DJIA a few days afterwards. Some other studies using Twitter to predict the stock market appeared in recent years. For instance, Rao and Srivastava (2012) combined Twitter sentiment with Google search volumes to predict returns, trading volume and volatility of commodities (e.g., oil, gold) and stocks. Sprenger et al. (2013) focus on tagged tweets (e.g., $MSFT representing Microsoft) and find a correlation of r=0.166 between Twitter sentiment and returns. Based on user posts from Twitter, online message boards as well as company news, Nann et al. (2013) created a trading model, which outperformed the S&P 500 index by 0.24 percent per trade after the consideration of transaction costs. Results from Oh and Sheng (2011), who study a three month period of roughly 70,000 postings on stocktwits.com, also reveal the predictive value of micro-blog messages to the stock market development.

Other social networks have also been investigated. Gilbert and Karahalios (2010) studied emotions extracted from LiveJournal, showing that the S&P 500 declines in case of increasing levels of anxiety. In a recent study, Karabulut

(2011) found that Facebook's Gross National Happiness (GNH) can predict returns in the US stock market.

In sum, studies from the offline as well as online world provide evidence that the stock market is driven by mood states of market participants. We therefore hypothesize:

> H1:   Increased social mood levels derived from Twitter lead to higher stock market returns.

There is also evidence in the literature that the community structure plays an important role when extracting mood from Social Media applications. Studies of diffusion processes and information cascades have a long tradition in the field of social network analysis as well as computer science (Granovetter 1973; Kempe et al. 2003; Leskovec et al. 2007; Hinz and Spann 2008).

We already know from experimental research that mood states are contagious (Hatfield et al. 1993). For instance, Bono and Ilies (2006) as well as Sy et al. (2005) found that followers and group members are influenced by positive mood states of their leaders. Neumann and Strack (2000) show that feelings are automatically transferred between individuals who listen to each other. Another example for emotional contagion in the real world comes from Fowler and Christakis (2008) who observed the spread of happiness in a real social network during a 20 year period.

A number of recent studies confirm these findings using an Internet setting. In the online world, it was shown that text-based communication can spread emotions among group members (Guillory et al. 2011; Hancock et al. 2008; Kramer 2012). Emotional contagion occurs on the Internet even in the absence of direct social interactions. In a recent experiment, Kramer et al. (2014) manipulated the volume of emotionally positive and negative posts in News Feeds of 689,003 Facebook users. It turned out that people who were exposed to less positive content produced fewer positive status updates themselves. On the other hand, if fewer negative posts occurred in their News Feeds, people published fewer negative status updates. The study confirms results of Coviello et al. (2014) who found that rainfall exerts an influence on the status messages of Facebook users as well as messages of geographically separated friends. Thus, emotions of Facebook members influence the emotions of other Facebook members. This relationship shows that textual content can spread emotions without direct social interactions.

The online shopping behavior also suggests that Internet users rely on the opinions of other community members. Conducting a field experiment, Grahl et al. (2014) were recently able to draw causal conclusions between social recommendations and purchase volume. Displaying Facebook Likes increases online

store revenues by almost 13 percent within one month, which indicates that Internet users are infected by opinions of their peers.

The Twitter network structure has also been investigated by previous research. So far, researchers only focused on the level of information or sentiment
spread but not on mood and emotional contagion. According to Lerman et al.
(2012), Twitter users are heavily connected with each other: Following friends
and re-tweeting messages leads to a large social network where news stories and
other content can easily spread. The authors previously presented a framework
for studying information cascades in online social networks (Ghosh and Lerman
2011). In general, using the number of re-tweets might be interesting for measuring emotional contagion. However, we realized that only a very small fraction of
tweets are re-tweeted. This observation is supported by empirical studies which
also found few re-tweets. For instance, Boyd et al. (2010) collected 720,000
tweets for studying the re-tweeting behavior on Twitter. Only 3 percent of the
tweets were re-tweets in this sample.

Oftentimes the number of Twitter followers has been used as a measure for
influence and popularity within the community (e.g., Cha et al. 2010; Kwak et al.
2010). The follower influence is also known as in-degree influence in the literature and describes the potential audience a user might reach (Bakshy et al. 2011;
Ye and Wu 2010). Bakshy et al. (2011) quantified user influence on Twitter and
concluded that, on average, "individuals who have been influential in the past
and who have many followers are indeed more likely to be influential in the
future". It is therefore reasonable to assume that the number of followers serves
as an appropriate measure for social influence (see for example Hinz et al. 2011).
Ruiz et al. (2012) study conversations about companies on Twitter and show
correlations with share prices under consideration of user activity and interaction
(e.g., number of followers, number of re-tweets). Hence, we hypothesize:

> H2:   Increased follower-weighted social mood levels derived from Twitter
>   lead to higher stock market returns.

4.3 Empirical Study

4.3.1   Data Collection and Method

We conduct our empirical study in three steps. First, we study a historical time
period in order to replicate previous studies investigating the relationship between Twitter mood and stock returns because it is unclear whether market actors already incorporate this new information and whether this market anomaly

still exists. We therefore collected tweets that have been published in Germany between January 1, 2011 and March 17, 2012. Afterwards, we integrate the number of followers into the analysis to see whether social interactions of Internet users help to predict the fluctuation of share prices. This second sample captures the period between December 1, 2012 and November 30, 2013.

We split this sample equally into a training period (December 1, 2012 - May 31, 2013) and a testing period (June 1 - November 30, 2013) in order to apply a trading strategy for investors. In the training period, we aim to investigate the predictive power of social mood states by integrating up to four lags into the model. Results of the trading strategy in the testing period will show whether investors might consider mood states for trades in the real world. We used a different time period for testing since applying a trading strategy in the same period would only reproduce existing results and therefore decrease the validity of the results (see for example Bollen et al. 2010 or Hill and Ready-Campbell 2011 who used a similar approach).

We accessed the data through the Twitter API[3]. Each tweet includes the tweet ID, time of publication, information on followers and re-tweets as well as text content, which is restricted to 140 characters. We eliminate all tweets that cannot be categorized as either positive or negative according to the dictionary approach described below.

For the mood analysis, we used Dalbert's (1992) "Aktuelle Stimmungsskala" (ASTS) which is the German version of the Profile of Mood States (POMS) originally developed by McNair et al. (1971). We therefore follow the seminal work of Bollen et al. (2010) who also used a modified version of POMS for extracting mood levels from public tweets. However, in contrast to these authors, we focus on one specific region (Germany) instead of collecting world-wide tweets.

The ASTS consists of 19 adjectives which belong to 5 mood dimensions: grief, hopelessness, tiredness, anger and positive mood. Respondents usually indicate on a 5-point scale how accurate each adjective describes their current feelings. For instance, the words hopelessly, discouraged and desperately are part of the hopelessness dimension. We expanded the original ASTS from 19 to 529 items by deriving synonyms from the German dictionary Wortschatz (Biemann et al. 2004). This larger scale is called WASTS (Table 4-1). We translated all items into English, which is the predominant language on Twitter. Only 1 percent of all Twitter messages are written in German, while 50 percent are written in English (Semiocast 2013). While the percentage of German tweets is higher in Germany, English is widely spoken on Twitter in this region (Leetaru et al. 2013). It is therefore reasonable to consider both English and German tweets

---

3   https://dev.twitter.com

Table 4-1: Depressive Mood States Derived by WASTS

ASTS dimension	Grief	Hopelessness	Tiredness	Anger	Positive Mood
WASTS dimension	Negative words (incl. synonyms)				Positive words (incl. synonyms)
Social Mood Index	Share of positive mood				

when measuring mood levels in Germany. However, it should be also clear that Twitter users in Germany are primarily German native speakers. According to Lewis (2009), more than 80 percent of all German native speakers are living in Germany. Emotional states are expressed differently across cultures and languages which differ widely in the size of their emotion lexicons (e.g., Benedict 1934; Boucher 1979; Brown and Gilman 1960; Gehm and Scherer 1988; Pavlenko 2008). Thus, using the English POMS scale for English tweets primarily written by Germans would ignore the cultural differences, which is why we translate the German WASTS scale into English (see also Gehm and Scherer 1988 for a similar approach).

Our approach enables us to classify tweets into one (or more) of the five WASTS mood dimensions. For instance, the tweet "I'm feeling good today" would increase the positive mood score by one point because of the occurrence of the word "good".

Our variable of interest for this study is the "Social Mood Index" (SMI), which we simply define as the share of positive mood on all word occurrences (sum of positive and negative mood states).

$$\text{Social Mood Index} = \frac{\text{Positive Mood}}{\text{Grief} + \text{Hopelessness} + \text{Tiredness} + \text{Anger} + \text{Positive Mood}} \tag{4.1}$$

That is, we sum up all positive and negative tweets each day in order to calculate SMI values. We used Central European Time (12 midnight) as cutoff time since we measured the social mood in Germany. The SMI is comparable to Facebook's Gross National Happiness (GNH) Index, which indicates the mood of Facebook users based on their status updates. The advantage of the SMI is that we do not have to rely on an external source (i.e. black box).

The SMI represents the Twitter mood in Germany. Every tweet, which has been published in Germany during our observation period, reflects a part of the social mood. One could argue that the social mood is not representative for the investors' mood. Indeed, if we were able to measure the investors' mood solely, we would expect an increase of accuracy of this assessment. However, not every investor has a public Twitter account and it is furthermore very difficult to identify all investors' nick names on Twitter. This is why we analyze the social mood on a macro level and assume that the overlap between social mood and investors' mood might be sufficient.

In this article, we follow Nofsinger (2005) who also used the term social mood for collective mood states[4]. He argues that "interaction with others has a strong influence and leads to a shared emotion, or social mood. Collectively shared opinions and beliefs shape individual decisions, which aggregate into social trends, fashion, and action" (pp. 8). According to this definition, the SMI is likely to capture a certain part of emotions of stock market participants.

In addition, we especially aim to consider the social character of mood states by integrating the number of followers into the analysis. The weighted social mood index (WSMI) is simply an extension of the original SMI in a way that we sum up all positive and negative mood followers each day:

$$Weighted\ Social\ Mood\ Index = \frac{Positive\ Mood*Followers}{Positive\ Mood*Followers+Negative\ Mood*Followers} \tag{4.2}$$

For instance, if an influential individual with 10,000 followers on Twitter is posting "I'm feeling good today", this positive tweet would increase the positive score by 10,000 points instead of one point (original SMI, see above).

Our dependent variable is the DAX intraday return, which we simply define as the percentage gain or loss between the first price and last price on the trading day. We then study whether SMI and WSMI values have predictive value to share returns. Most of the previous studies have found a relationship between shifts in mood states and a stock market reaction on the next trading day (see section 4.2). For instance, Kamstra et al. (2000) show that time changes on Sunday ("daylight saving anomaly") leads to negative abnormal returns on the following Monday. Edmans et al. (2007) found a negative stock market reaction on the trading day after the elimination of the national soccer team at the World Cup. According to Karabulut (2011), changes of Facebook's Gross National Happiness predict S&P 500 changes on the next trading day. However, Bollen et al. (2010) found significant values for different lags so that we take this possibility into account by including more than one lag into the analysis. This is especially interesting when investigating emotional contagion effects (H2).

It should be noted that the DAX is dominated by foreign investors. However, these investors are mostly institutional investors such as banks or insurance companies which should not be prone to sentiment changes. For instance, the world's biggest money manager Black-Rock owns four percent of DAX total

---

4   See also background mood (Loewenstein et al. 2001).

value[5]. In contrast, individual investors and therefore noise traders are mostly domestic investors, living in Germany in our case (see section 4.2 for a discussion on noise traders). The preference for domestic stocks is known as "home bias" in the literature (French and Poterba 1991). The reason why retail investors prefer local stocks might be familiarity (Huberman 2001; Grinblatt and Keloharju 2001) or superior information (Coval and Moskowitz 1999). We therefore assume that a visible stock market reaction can be observed if noise investors, who are primarily German retail investors, are affected by changing mood levels which in turn influence their risk-taking tendencies.

Equation 4.3 depicts that we control for a number of anomalies, which have been discussed in the previous research section. We account for technical related anomalies by the DAX intraday performance on the previous day (r_{t-1}). This momentum variable represents the general market development (bull or bear market). The DAX index consists of 30 major German companies. It has been shown that past winners are often future winners and vice versa (Chan et al. 2000). In addition, we control for calendar anomalies (see section 4.2.1). Therefore we integrate dummy variables for trading days after the weekend ($Monday_t$) and national holidays. Further, the tax dummy variable equals 1 for December 28, 2012 (last trading day of the tax year) as well as January 2-8, 2013 (first five trading days of the tax year) in order to account for tax-loss selling. We take the lunar cycle into account (Dichev and Janes 2003) by constructing a dummy variable which equals 1 for the (-3;+3) window around full moon days and 0 otherwise. Finally, we control for a time trend by including $Time_t$. This variable equals 1 on the first trading day of the observation period, 2 on the second trading day and so forth.

We also control for investor sentiment proxies: trading volume, stock market volatility and consumer confidence. Trading volume and volatility have been shown to interact with stock indices in the past (e.g., Chen et al. 2001; Chordia and Swaminathan 2000; French et al. 1987; Gallant et al. 1992; Karpoff 1987). $TradingVolume_t$ represents the turnover of all DAX shares on day t. $Volatility_t$ is the stock market volatility on day t as measured by the VDAX-NEW. This index indicates the implied volatility of the DAX which is expected by market participants for the next 30 days[6]. In addition, we include the consumer confidence into the analysis. Qiu and Welch (2006) have shown that consumer sentiment correlates well with investor sentiment. Furthermore, Lemmon and Portniaguina (2006) used consumer confidence as measure for investor sentiment in order to forecast share returns. One prominent measure for consumer confidence

---

5   See Germany Trade & Invest for more information on foreign investors: http://www. gtai.de/GTAI/Navigation/EN/Invest/Service/Publications/Markets-germany/Archive/ Issues-2011/Volume-2/Fdi/foreign-investors-put-faith-germanys-stocks.html
6   More information on VDAX-NEW can be found at the website of the exchange:

in Germany is the GfK index, which is published by the market research group GfK once a month[7]. $ConsumerConfidence_t$ indicates the consumer confidence as measured by the GfK index (in points) in the respective month on day t. We use OLS in order to measure the effect of Twitter mood on stock returns. We estimate our model (equation 4.3) with robust standard errors due to heteroskedasticity (Breusch-Pagan test p<.01).

$$r_t = \beta_0 + \beta_1 * SMI_{t-1} + \beta_2 * SMI_{t-2} + \beta_3 * SMI_{t-3} + \beta_4 * SMI_{t-4} + \beta_5 * r_{t-1}$$
$$+ \beta_6 * TradingVolume_t + \beta_7 * Volatility_t + \beta_8 * ConsumerConfidence_t$$
$$+ \beta_9 * Monday_t + \beta_{10} * Holiday_t + \beta_{11} * Tax_t + \beta_{12} * Moon_t + \beta_{13} * Time_t + e_t$$

$$(4.3)$$

4.4 Results

4.4.1 Descriptive Statistics

In our historical sample, we observe the highest SMI value (0.679) on January 1 2012, while the Twitter mood was rather low when Amy Winehouse died (0.617) or during a terrorist attack in Moscow on January 24, 2011 (0.602). It should be noted that we do not aim to show a causal relationship between these events and share returns in this article. As described in section 4.2, mood states can be influenced by many factors, such as stress levels, weather conditions, social interactions, etc.

Overall, the historical sample period contains 310 trading days between January 1, 2011 and March 17, 2012. The mean value of the SMI during this period is 0.637, which means that two third of tweets were recognized as being positive. The phenomenon that positive words are used more often than negative words is known as "Pollyanna effect" in the literature (e.g., Boucher and Osgood 1969).

This number fits well with previous studies extracting sentiment from Internet messages. For instance, Rao and Srivastava (2012) studied stock and commodity discussions on Twitter and found that 67.14 percent of tweets were positive. The ratio between positive and negative tweets persists when calculating WSMI values. Figure 4-2 in the appendix shows a comparison between the WSMI and SMI over time. Overall, we collected roughly 100 million tweets in the three years period between January 2011 and November 2013. On average, 102,084 tweets per month are recognized by the German and English version of

---

7   Information on GfK index can be found on http://www.gfk.com.

the ASTS scale. While 60 percent of tweets are English, 40 percent are recognized as German tweets.

4.4.2   Relationship between Social Mood and the Stock Market

We surprisingly find no significant relationship between Twitter mood as measured by the SMI and share returns on the next 4 trading days in Germany (Table 4-2). We can therefore reject Hypothesis 1. One explanation might be that market actors have incorporated the mood level in their models so that the market anomaly is not persistent anymore. Multicollinearity does not seem to be a problem with all VIFs below 10 (mean VIF = 1.57).

Table 4-2: Influence of SMI on the Stock Market (01/2011 – 03/2012)

	Coefficient	Robust Std. Err.	t-value	P>t
Constant	.078	.060	1.32	.188
SMI_{t-1}	-.034	.065	-.052	.602
SMI_{t-2}	-.007	.076	-0.09	.925
SMI_{t-3}	-.068	.076	-0.90	.369
SMI_{t-4}	.027	.076	0.36	.719
r_{t-1}	-.036	.071	-0.52	.607
Trading Volume	-.002	.001	-1.61	.107
Volatility	-.000	.000	-0.77	.442
ConsumerConfidence	-.002	.005	-0.48	.634
Calendar controls	Yes			
Time Control	Yes			
Number of observations: 310; R²: 0.032; Mean VIF: 1.57				

4.4.3   Relationship between Follower-Weighted Social Mood and the Stock Market

Previous research has shown that mood states and emotions are contagious on the Internet (e.g., Kramer et al. 2014). We also know that Internet users heavily interact with each other on micro-blogs. It is therefore reasonable to investigate whether the predictive ability of the SMI improves when weighting each tweet according to its importance within the Twitter atmosphere. We therefore include the number of followers into the analysis and create the WSMI as described in section 4.3.

Table 4-3:   Influence of WSMI on the Stock Market (12/2012 – 05/2013)

	Coefficient	Robust Std. Err.	t-value	P>t
Constant	-.106*	.059	-1.80	.075
$WSMI_{t-1}$	.033**	.016	2.09	.039
$WSMI_{t-2}$	.011	.012	0.88	.382
$WSMI_{t-3}$	-.019	.017	-1.15	.252
$WSMI_{t-4}$	.014	.021	0.64	.522
r_{t-1}	-.075	.099	-0.76	.446
Trading Volume	-.001	.001	-0.69	.491
Volatility	-.002***	.001	-4.18	.000
ConsumerConfidence	.021**	.009	2.25	.026
Calendar controls	Yes			
Time Control	Yes			

*** Significant at the 1 percent level; ** Significant at the 5 percent level; * Significant at the 10 percent level; Number of observations: 117; R^2: 0.25; Mean VIF: 2.07

Please note that this information is not available for the historical data set that we used in our first analysis. Our second sample includes tweets that have been published between December 1, 2012 and May 31, 2013 in Germany. We study the influence of the WSMI on the stock market on 117 trading days.

Table 4-3 shows that the DAX intraday return is positively influenced by increased WSMI values, supporting H2 ($p<.05$). A one percent increase of the WSMI compared to the previous day exerts an influence of 3.3 basis points on the next day's DAX return[8]. The relatively small effect of 3.3 basis points fits well with existing studies, which investigated the predictive value of mood states and online sentiment to the stock market. Most researchers observe only weak magnitudes (e.g., Antweiler and Frank 2004; Karabulut 2011). Compared to other studies in the field of share price forecasting, our R^2 value of 25 percent is relatively high[9]. Usually small R^2 values are reported due to the fact that share prices are influenced by a number of factors, which cannot be included into one regression. Even the R^2 of 3.2 percent which we receive in the historical sample (Table 4-2) is at the upper end of existing studies. As a robustness check, we also calculated (W)SMI values without the anger dimension due to the fact that anger

---

8   We also calculated SMI and WSMI values without the anger dimension and receive qualitatively similar results.

9   Among others, Antweiler and Frank (2004) report R^2 value of 0.049; Avery et al. (2009) report R^2 values between 0.0005 and 0.0151; Das and Chen (2007) report R^2 value of 0.0027 and 0.0041.

might foster risk-taking tendencies and thus lead to higher stock market returns. However, we receive qualitatively similar results compared to our original SMI and WSMI measures (see appendix, Tables 4-7 to 4-9).

We found only one working paper, which included the number of followers into the Twitter mood analysis. In contrast to our results, Zhang et al. (2010) do not report any significant influence of follower-weighted mood levels on the US stock market. However, the authors only present correlation coefficients of Twitter mood variables with the US stock market and do not perform more sophisticated analyses or control for other mood and technical-related anomalies.

We adopted a bivariate VAR model in order to test Granger causality. The model is estimated with the following equation:

$$z_t = \alpha + \sum_{j=1}^{n} \gamma_j * z_{t-j} + \beta * x_t + e_t \qquad (4.4)$$

where z_t is a vector of the WSMI and DAX intraday return on day t. x_t is a vector of our control variables: $TradingVolume_t$, $Volatility_t$, $ConsumerConfidence_t$, $Monday_t$, $Holiday_t$, Tax_t, $Moon_t$, $Time_t$.

In contrast to OLS regression, the VAR model allows to capture linear interdependencies among the follower-weighted social mood and share returns. That is, the variables are explained in the VAR system both by their own delayed values as well as by the delayed values of the other variable. Testing up to 10 lags, we receive the lowest Akaike Information Criterion (AIC) and Schwarz' Bayesian Information Criterion (BIC) when choosing 1 lag (=1 day). The WSMI exerts a significant influence (p<.05) on the next day's DAX return (see Table 4-4). Furthermore, the granger causality test shows that the WSMI does actually granger-cause the DAX intraday return (p<.05).

Table 4-4:   Results of VAR model (December 1, 2012 – May 31, 2013)

Dependent variable: DAX intraday return (r_t)	Coefficient	Std. Err.	t-value	P>t
Constant	-.096*	.055	-1.74	.082
Lag 1 WSMI	.032**	.016	1.98	.048
Lag 1 DAX intraday return	-.083	.086	-0.97	.332
TradingVolume	-.001	.001	-0.61	.539
Volatility	-.003***	.001	-4.92	.000
ConsumerConfidence	.020**	.009	2.10	.036
Calendar controls	Yes			
Time Control	Yes			
*** Significant at the 1 percent level; ** Significant at the 5 percent level; * Significant at the 10 percent level; R²: 0.23				

The unweighted SMI variable, which measures the social mood without the consideration of follower numbers, is again far from being significant in the sample period (see Table 4-6, appendix).

4.5 Trading Strategy

Based on our results, we created a virtual portfolio and applied a simple trading strategy. Individuals can easily invest in stock indices with the help of exchange-traded funds (ETF). These highly liquid funds can be bought and sold during regular trading hours and fully replicate the index performance. If the WSMI increases compared to the previous day, we buy the iShares DAX ETF (ISIN DE0005933931), which is the most popular ETF in the German market. We then hold the investment for one trading day so that our win or loss is the difference between the last price and the first price of the focal trading day. In case of decreasing WSMI values, we buy the db x-trackers ShortDAX ETF (ISIN LU0292106241), which is a liquid instrument in order to benefit from decreasing DAX values.

The trading strategy is applied to a time period (June 1 - November 30, 2013) different from the training period in order to test whether there actually is a predictive value associated with social mood. Again, only tweets that have been identified as being relevant by our dictionary were stored in the database. The WSMI was calculated in the same way as described in section 4.3.1.

The following example illustrates our approach: The WSMI decreased from 0.75 points on Wednesday, June 19 to 0.71 points on Thursday, June 20. We then buy the ShortDAX ETF on Friday, 21 June. On this day, the DAX decreased by 1.98 percent from 7946.32 points (first price in the morning) to 7789.24 points (last price in the evening). The ShortDAX ETF increased by 1.91 percent so that the portfolio has realized a win of roughly 2.0 percent before transaction costs. These numbers illustrate that long as well as short ETFs replicate the index performance virtually on a 1:1 ratio. We chose two highly liquid ETFs in order to create a realistic investment scenario. However, investors are not restricted to these ETFs and might use other instruments.

Starting with € 100,000 on June 1, 2013, this portfolio would increase to € 121,012 until the end of our observation period on November 30, 2013 (Table 4-5). Thus, this simple trading strategy delivers a return of more than 20 percent within 6 months while the DAX index itself only increased by 13.4 percent (see also P&L chart in Figure 4-3, appendix).

Table 4-5:   Trading Strategy

	Performance (June 2013-November 2013)	
WSMI Strategy (Long/Short)	Before Transaction Costs	After Transaction Costs
DAX ETF (1:1 ratio)	21.01%	19.11%
Leveraged DAX ETF (2:1 ratio)	37.56%	35.63%
Benchmark indices		
DAX	13.40%	
Euro Stoxx 50	11.93%	
S&P 500	10.67%	

The outperformance against the DAX persists even if we control for trans-action costs. Assuming a brokerage fee of € 5 per trade[10], transaction costs would reduce the return of the portfolio by € 10 each day. However, this trading strategy would still realize a positive six-month performance of 19.11 percent, increasing the value of the portfolio from € 100,000 to € 119,114.

It can further be improved by investing into leveraged ETFs. These funds are also easy to buy, tracking the index performance on a ratio of for example 2:1 or 3:1. We use the db x-trackers LevDAX ETF (ISIN LU0411075376) for long investments and the db x-trackers ShortDAX x2 (ISIN LU0411075020) in order to benefit from decreasing DAX values. The 2x leveraged ETF strategy would achieve a return of 35.63 percent after transaction costs.

Next, we calculate the Sharpe Ratio, which is a common reward-to-volatility measure (Sharpe 1966):

$$\text{Sharpe Ratio} = \frac{(R_a - R_b)}{\sigma} \tag{4.5}$$

where R_a represents the return of an asset (DAX return in our case); R_b denotes the return of the riskless investment as measured by the risk-free interest rate; σ represents the standard deviation of the excess returns $(R_a - R_b)$.

The Sharpe Ratio determines the return per unit of risk. Assuming 260 trading days, the average daily return in our case is 0.164 percent or 42.60 percent on an annual basis. If we further deduct the risk-free interest rate of 3 percent, which is close to the long-term mean value (e.g., Hill and Ready-Campbell 2011), we receive an excess return of 39.60 percent. The standard deviation of

---

10 There are several discount brokers offering their clients cost-effective access to capital markets (e.g., Cortal Consors in Germany). We are aware that € 5 is at the low end of the range. However, these costs are very easy to realize for the individual investor. Nevertheless, the outperformance against the benchmark indices would persist even if we assume € 10 per trade.

daily returns is 0.0016 or 0.104 annualized. Thus, the Sharpe Ratio of the trading strategy is 3.8, which means that the investor is compensated well for the risk taken.

Despite this promising performance, we are aware that there are usually other transaction costs in addition to the brokerage fee. The bid/ask spread might be severe, especially for less liquid investment products. However, this spread is virtually zero for DAX ETFs due to large turnover rates and the great competition among market makers. Operating expenses (i.e. costs for administration, portfolio management, etc.) are another part of transaction costs. However, these are very low for ETFs since there is no portfolio management in contrast to actively managed funds. For instance, the total expense ratio of the iShares DAX ETF is only 0.17 percent per year. In sum, we are confident that investors can use social mood states for their investment success, even after the consideration of transaction costs.

4.6 Conclusion

Our results provide evidence that follower-weighted social mood levels can predict share returns. An improved WSMI of one percent leads to a 3.3 basis points DAX increase on the next trading day during our training period. This effect is persistent even if we control for other anomalies, such as calendar effects. Surprisingly, our results do not support the view that the simple aggregation of mood states of all individuals in the Twitter blogosphere is enough to predict the stock market. It is rather necessary to consider the community structure (i.e. followers). An explanation for this phenomenon might be emotional contagion among Internet users as has been shown by previous research (e.g., Kramer et al. 2014).

The missing effect of the non-weighted SMI might be explained by the fact that some investors already conduct data mining and collect messages from Social Media applications in order to buy or sell stocks according to mood levels. Mood analysis is increasingly gaining interest and a number of companies emerged in recent years, offering their clients solutions to analyze big data on the Internet. Previous research used Twitter and Facebook data primarily from the years between 2007 and 2011 (e.g., Bollen et al. 2010; Karabulut 2011). Meanwhile, many articles were published by academic journals and the media so that investors are more likely to be aware of the large potential of user-generated content on the Internet. Our sample covers a more recent time period between 2011 and 2013. Thus, while previous research used social mood states primarily as private data (i.e. not visible for most investors), Twitter mood could be public data by now (i.e. visible for many or large investors), making financial markets more efficient and decreasing the predictive value of Social Media applications. The diminishing influence of Twitter messages on the stock market might be compared with other mood-related anomalies, such as the weather effect. Saunders (1993) presented evidence for a sunshine effect in the US stock market during a 100 year period, although results in the last period (1983-1989) have not been statistically significant. In addition, researchers tried to reproduce Saunders' study in subsequent years but many of them could not find a significant relationship between weather conditions and share prices (e.g., Krämer and Runde 1997; Trombley 1997; Worthington 2009). This lack of significance might be the product of data mining strategies, which make financial markets more efficient over the years. Our study may potentially indicate similar effects for mood states derived from Social Media applications, although we can only speculate at this point in time. However, one has to be careful when interpreting these results. The insignificance of the SMI might be also caused by our measurement. We are confident that the German and English version of the WASTS scale is best suitable to assess mood states of the German Twitter users. It might be problematic to use the English POMS scale to assess mood states of German native speakers due to cultural differences in emotion lexicons (e.g., Pavlenko 2008). Nevertheless, it was used for the first time when studying the influence of mood states on share returns. The WASTS deviates to some extent from other scales previously used by researchers who found significant mood effects (e.g., Bollen et al. 2010).

The consideration of social interactions among community members delivers promising results. Follower-weighted social mood states have predictive value to stock returns. Our simple trading strategy, which we applied for the German stock market, delivers returns between 19.11 percent and 35.63 percent after the consideration of transaction costs. We are therefore able to outperform major international benchmark indices by double-digit percentage points.

Our results have strong implications for investors as well as the entire economy. The financial industry might integrate mood levels into traditional forecast models to make better trading decisions. Especially the combination of mood analysis with established capital market models would be an interesting area for future research in order to further improve forecast accuracy.

Implications of our results are not restricted to the financial industry. Future research might also investigate the relationship between social mood levels and other areas of our economy. For instance, the buying behavior of consumers seems to be influenced by emotions and feelings (Weinberg and Gottwald 1982). Researchers might predict online sales with the help of social mood levels derived from Twitter or Facebook.

Our results might be the first indication that emotional contagion caused by online messages can influence people's behavior in the offline world, particularly the economic behavior. It therefore might be possible for Facebook, Twitter or another massive social network to manipulate the amount of positive messages shown to users in order to improve the economy. However, we cannot actually prove emotional contagion at this point in time. We can only assume the spread of mood states among Twitter users. Although there is evidence for emotional contagion on the Internet and Facebook in particular (e.g., Coviello et al. 2014; Kramer et al. 2014), the magnitude of mood transfers on Twitter could be identified by future research projects. Another avenue for future research would be to study intraday instead of inter-day effects of mood swings. There is already some evidence that shifts of investors' mood states can influence share prices during the trading day (e.g., Chang et al. 2008; Lo and Repin 2002) and it would be interesting to study the influence of intraday mood swings derived from Twitter or Facebook. In addition, researchers could integrate other Internet sources, such as discussion boards or news sites. Especially the consideration of market news would help to compare the influence of mood states with the influence of events which occur in the real world.

Despite our promising results, our research has still some shortcomings. There may be fake messages in our sample. However, according to Twitter, only 5 percent of all accounts are fake (D'Onfro 2013). Studies focusing on the predictive value of Twitter also found similar numbers of spam accounts (e.g., Conover et al. 2011). It is furthermore questionable whether these accounts actually produce fake messages which potentially pose a threat to the validity of our research.

Our dictionary approach does not consider specific features of tweets, such as emoticons and Internet slangs (e.g., Bifet and Frank 2010). These features might also convey mood, which is currently not captured by our SMI and WSMI. Our dataset for studying the influence of follower-weighted mood states is relatively small. Overall, it captures the one-year period between December 1, 2012 and November 30, 2013. Further analyses with larger datasets are required in order to confirm our results. Especially changing market phases might deliver different results of our trading strategy. We used different time periods for training and testing and therefore followed Bollen et al. (2010) as well as other authors who used data of Social Media applications (e.g., Hill and Ready-Campbell 2011). However, several researchers (e.g., Ali and Pazzani 1992; Holte et al. 1989) argue that using different market phases for training and testing might cause incorrect results due to the problem of "small disjuncts". Therefore it might be interesting to apply our trading strategy in the real world in order to test the validity of the results. Sentiment and mood analysis with the help of Social Media is still a relatively young research domain. However, academia and industry are more and more aware of the huge potential for predicting the company success. It is difficult to evaluate how mood analysis will change the financial industry. According to our results, the network structure should be considered when studying the relationship between mood levels and share returns. In sum, opportunities in the field of mood analysis seem to be unlimited for researchers and practitioners which is why we have to expect numerous research projects over the next few years.

4.7 Appendix

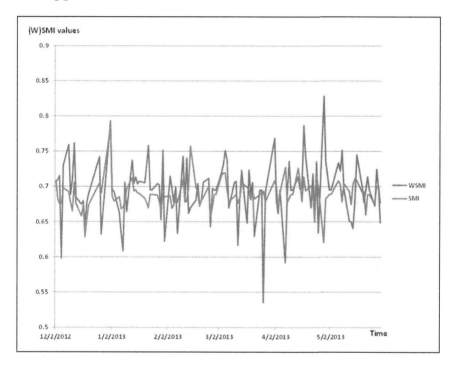

Figure 4-2: SMI and WSMI Values over Time

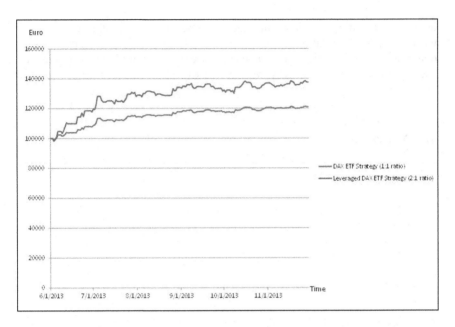

Figure 4-3:   P&L Chart of Trading Strategies between June 1, 2013 and November 30, 2013

Table 4-6:   Influence of SMI on the Stock Market (12/2012 – 05/2013)

	Coefficient	Robust Std. Err.	t-value	P>t
Constant	-.130	.086	-1.52	.133
SMI_{t-1}	.018	.037	0.48	.635
SMI_{t-2}	.015	.032	0.48	.631
SMI_{t-3}	.039	.046	0.86	.391
SMI_{t-4}	-.012	.040	-0.30	.766
r_{t-1}	-.074	.099	-0.74	.459
Trading Volume	-.000	.001	-0.15	.880
Volatility	-.003***	.001	-4.14	.000
ConsumerConfidence	.022**	.009	2.55	.012
Calendar controls	Yes			
Time Control	Yes			
*** Significant at the 1 percent level; ** Significant at the 5 percent level; Number of observations: 117; R²: 0.21; Mean VIF: 2.37				

Table 4-7: Influence of SMI without Anger on Share Returns (01/2011 – 03/2012)

	Coefficient	Robust Std. Err.	t-value	P>t
Constant	.082	.056	1.47	.143
SMI_{t-1}	-.006	.063	-0.10	.924
SMI_{t-2}	-.023	.072	-0.32	.752
SMI_{t-3}	-.068	.075	-0.91	.364
SMI_{t-4}	.009	.073	0.12	.905
r_{t-1}	-.036	.071	-0.51	.608
Trading Volume	-.002	.001	-1.65	.100
Volatility	-.001	.000	-0.80	.426
ConsumerConfidence	-.002	.004	-0.48	.634
Calendar controls	Yes			
Time Control	Yes			
Number of observations: 310; R²: 0.032; Mean VIF: 1.59				

Table 4-8: Influence of WSMI without Anger on Share Returns (12/2012 – 05/2013)

	Coefficient	Robust Std. Err.	t-value	P>t
Constant	-.120*	.061	-1.96	.053
$WSMI_{t-1}$	.033**	.016	2.02	.046
$WSMI_{t-2}$	.019	.013	1.43	.156
$WSMI_{t-3}$	.011	.017	-0.65	.520
$WSMI_{t-4}$	.022	.022	0.98	.329
r_{t-1}	-.080	.097	-.083	.409
Trading Volume	-.001	.001	-.053	.597
Volatility	-.002***	.001	-4.21	.000
ConsumerConfidence	.020**	.009	2.20	.030
Calendar controls	Yes			
Time Control	Yes			
***Significant at the 1 percent level; **Significant at the 5 percent level; * Significant at the 10 percent level; Number of observations: 117; R²: 0.25; Mean VIF: 2.06				

Table 4-9:   Influence of SMI without Anger on Share Returns (12/2012 – 05/2013)

	Coefficient	Robust Std. Err.	t-value	P>t
Constant	-.142	.086	-1.64	.104
SMI_{t-1}	.018	.032	0.55	.584
SMI_{t-2}	.024	.027	0.92	.359
SMI_{t-3}	.032	.037	0.85	.396
SMI_{t-4}	-.007	.032	-0.20	.838
r_{t-1}	-.079	.099	-0.80	.428
Trading Volume	-.000	.001	-0.06	.949
Volatility	-.003***	.001	-4.06	.000
ConsumerConfidence	.024**	.009	2.58	.011
Calendar controls	Yes			
Time Control	Yes			

*** Significant at the 1 percent level; ** Significant at the 5 percent level; Number of observations: 117; R²: 0.22; Mean VIF: 2.41

5 The Economic Impact of Privacy Violations and Security Breaches – A Laboratory Experiment[1]

Abstract

Privacy and security incidents represent a serious threat for a company's business success. While previous research in this area mainly investigated second-order effects (e.g., capital market reactions to privacy or security incidents), this study focuses on first-order effects, that is, the direct consumer reaction. In a laboratory experiment, the authors distinguish between the impact of privacy violations and security breaches on the subjects' trust and behavior. They provide evidence for the so-called "privacy paradox" which describes that people's intentions, with regard to privacy, differ from their actual behavior. While privacy is of prime importance for building trust, the actual behavior is affected less and customers value security higher when it comes to actual decision making. According to the results, consumers' privacy related intention-behavior gap persists after the privacy breach occurred.

5.1 Introduction

A series of cyber-attacks in recent years at global companies like Sony, Citigroup, Lockheed Martin, Google, and Apple have shown that even large companies are vulnerable to attacks that threaten the protection of their costumers' data. Most recently, 250,000 Twitter accounts (Kelly 2013) and up to 6.5 million LinkedIn user accounts have been hacked (Silveira 2012). These security incidents can lead to serious consequences for the affected companies. For instance, Sony had to close their PlayStation network and their Online Entertainment platform for several weeks in May 2011 after hackers had been able to get access to 77 million user accounts, extracting customer information such as passwords, home addresses, and dates of birth (Bilton and Stelter 2011). As a result, the

---

1 Nofer, Michael / Hinz, Oliver / Muntermann, Jan / Roßnagel, Heiko (2014). The Economic Impact of Privacy Violations and Security Breaches – A Laboratory Experiment. Business & Information Systems Engineering, forthcoming.

company spent USD 170 million to cover the costs for increased customer support, data security improvements, and overall investigations into the incident.

In the long run, indirect consequences might be an even bigger threat to company success. Since privacy was identified as a major antecedent of trust, the relationship between existing and prospective clients and the company may permanently suffer. Several attempts to study the link between privacy, trust, and the intention to buy a product have been reported in literature, especially in the e-commerce environment, where trust plays an important role for business (Eastlick et al. 2006; Gefen 2000; Kim et al. 2008; Liu et al. 2005). These studies suggest a direct connection between privacy, security, and the buying intention, as well as a strong impact of privacy and security on trust in the company, which in turn influences the willingness to enter a business relationship.

Determining the impact of privacy violations and security breaches in monetary terms is quite challenging. This is due to the various factors that affect company success, so that the influence of privacy and security cannot easily be isolated from other effects. The event study methodology is often used to assess the economic impact of privacy and security incidents (Acquisti et al. 2006; Andoh-Baidoo et al. 2010; Cavusoglu et al. 2004; MacKinlay 1997). However, this approach is based on the strong assumption that the market correctly and fully reflects the impact of the event (e.g., security breach) on the customers' behavior and that the effect can be isolated from other effects.

Against this background, the motivation of our study is to explore the causal effect of data protection violations on consumer behavior by conducting a laboratory experiment. The goal is to analyze and compare the economic impact of privacy violations and security breaches. For this purpose, we use one control and two treatment groups. Whilst no data protection problem occurs in the control group, the other two groups are confronted with a privacy and respectively a security incident of a fictional bank. We first provide general information about the bank (see Appendix). For this we use information on one of the largest European banks from Wikipedia which we slightly adapted (e.g., changed the name). This description also includes information on a) a privacy violation in the recent past, b) a security breach in the recent past or c) none of these incidents. After this short description of the bank's characteristics, subjects were informed of the investment conditions of this bank, which is identical for all three conditions. The subjects then have to decide how much of their own money they are willing to invest in a financial product offered by the fictional bank. The money invested can also be lost with a probability that is identical for all three scenarios. It is important to note that subjects are not aware of the other scenarios but only get the information for the group to which they were randomly assigned (see section 5.5.1 for details).

We adapt the economic decision game called the "investment game", first introduced by Berg et al. (1995), in a way that allows us to compare the proportion of investments between the groups, thereby isolating the impact of security breaches and privacy violations, since all the other information on bank characteristics and investment conditions are identical for all participants. Beside this monetary impact, we also investigate how trust in the bank is affected and how trust in turn influences the willingness to invest. Thus, we can determine and compare the direct and indirect impact of privacy violations and security breaches on the investment amount. Many other studies investigate privacy and security issues from the viewpoint of the capital market and show the influence on share prices (Acquisti et al. 2006; Andoh-Baidoo et al. 2010; Cavusoglu et al. 2004). The stock market reflects the investors' expectations with regard to the company's future success. In contrast to these second-order effects, our study focuses on first-order effects, that is, the direct customer reaction to privacy violations and security breaches and thus offers a new way to quantify the impact of privacy and security issues. In addition, as a subordinate research goal, we aim to answer the question whether the so-called "privacy paradox" persists after a privacy breach occurred. The privacy paradox was demonstrated by researchers and means that consumers do not act according to their stated privacy concerns (e.g., Berendt et al. 2005; Dommeyer and Gross 2003; Norberg et al. 2007). So far, consumer behavior was studied without the occurrence of privacy or security incidents. We can therefore extend previous findings and test whether consumers change their behavior after a company suffers privacy breaches.

We first refer to work related to our study and then discuss the concepts of privacy and security, as well as their close link to trust and behavioral intentions. We present previous findings, emphasizing the meaning of trust for relationships and business activity in particular. We proceed with our research model and hypotheses, before we present the empirical results. We conclude the article with a discussion of the findings and ideas for future research.

5.2 Related Work

Following the own privacy policy is crucial for companies. Culnan and Armstrong (1999) show that fair behavior can build trust and that retention rates will be higher if clients perceive to be treated fairly. Thus, companies should behave in line with their rules, which should be externally communicated in order to increase the likelihood of obtaining personal information from consumers. In contrast, John et al. (2011) found that disclosing the own privacy policy and informing about data protection can actually lower consumers' willingness to provide personal information since privacy concerns increase. However, Hinz et al. (2011) found that honestly revealing the use of data can increase profits. This is also confirmed by Tsai et al. (2011) who show that the display of privacy policies positively influences the purchase intention and consumers even pay a price premium for more privacy protection.

Privacy violations also affect the company's reputation, a critical factor for long-term success. In a literature review, Yoon et al. (1993) report various findings about the role of company reputation and show empirically that the company's reputation has a direct and indirect impact on the intention to buy a product.

The impact of privacy violations and security breaches on a firm's value has been addressed by a number of empirical analyses on the basis of the event study methodology. Here, authors measure excess stock market returns of listed firms that have been affected by a corresponding event. Andoh-Baidoo et al. (2010) for example observe the impact of security breaches that have been reported in major US newspapers. They detect significant stock price reactions within an event period of three days starting one day prior to the event date. In contrast, Acquisti et al. (2006) address the impact of privacy violations on a firm's market value. Their results provide evidence for a significant but moderate price effect that can be observed during the two days subsequent to the publication. While event studies are well-recognized in empirical research, there exist a number of possible biases that can affect results (Campbell et al. 1997). One major problem results from uncertainty about the event dates when collecting them from financial publications. Other problems can result from non-trading or non-synchronous trading that for example occurs due to the fact that used closing prices do not have a common timestamp since they result from the last transaction of a trading day. As noted by Acquisti et al. (2006), limitations can also arise due to small sample sizes that would also be needed to "understand and contrast the impact of 'pure' security breaches compared to privacy ones", which also provides motivation for future research and to "study empirically the implications of privacy violations that go beyond their stock market influence" (p. 1579).

5.3 Theoretical Background

5.3.1 Privacy

There is no consistent definition of privacy and many researchers see it as a multidimensional construct (Foxman and Kilcoyne 1993; Goodwin 1991; Prosser 1960). The ambiguousness may be due to the different areas where the concept of privacy is used and discussed. Generally, one can distinguish between physical privacy and information privacy (Smith et al. 2011). The former refers to an individual's ability to live undisturbed and without interferences within private surroundings. Information privacy has increasingly gained in importance since the beginning of the information age. Unless otherwise stated, we use privacy as a synonym for information privacy. One popular notion in political science comes from Westin (1967), specifying privacy as "the ability of individuals to control the terms under which their personal information is acquired and used".

The element of control is especially important for the relationship between companies and consumers due to the increasing collection of personal information in recent years. Goodwin (1991) defines consumer information privacy as "the consumer's ability to control (a) presence of other people in the environment during a market transaction or consumption behavior and (b) dissemination of information related to or provided during such transactions or behaviors to those who were not present".

We furthermore refer to Greenaway and Chan (2005) who make a distinction between consumer information privacy and organizational privacy which describes "how firms treat their customers' personally identifiable information". The simulated privacy breach, which is described below, affects consumers' privacy but is also a case of organizational privacy due to the unfair treatment of consumer information by the company.

Researchers have repeatedly shown that consumers are concerned about privacy and the way that companies treat their personal information (e.g., Berendt et al. 2005; Phelps et al. 2000). Since the end of the 20th century, advances in information technology make it easier for companies to collect and distribute information. Therefore privacy concerns emerge especially with regard to the secondary use of personal information (Culnan 1993). Unauthorized secondary use exists when data is collected for one purpose but used for another purpose without the individual's permission. Smith et al. (1996) identified three other dimensions being central to the individual's privacy concerns. The collection of personal information reflects the fear that too much data about the individual is collected in society. Another area of concern is the improper access, which means that people within the organization have unjustifiable access to the customer information. The fourth dimension is an error in personal data, which might result from typing errors or accidental mistakes. Consistent across cultures, unauthorized secondary use of information was found to be the most important concern dimension for consumers (Milberg et al. 1995).

5.3.2 Security

For the purpose of our research, it is important to make a distinction between privacy and security, although some authors use these concepts interchangeably or summarize the concepts under new terms, such as "structural assurance" (Luo et al. 2010; McKnight and Chervany 2001).

Security concerns increased significantly since transactions can be done over the Internet. Recent cyber-attacks at Sony or Citigroup show the vulnerability of today's technology. Consumers are afraid of criminal activities, such as information theft and data fraud (Suh and Han 2003). This is why many studies identified perceived security as a major antecedent of consumers' willingness to purchase from e-commerce stores (Belanger et al. 2002).

Kalakota and Whinston (1996) define a security threat as a "circumstance, condition, or event with the potential to cause economic hardship to data or network resources in the form of destruction, disclosure, modification of data, denial of service, and/or fraud, waste, and abuse."

Smith et al. (2011) review more than 300 privacy articles and differentiate between privacy and security in such a way that security concerns result from concerns about: "integrity that assures information is not altered during transit and storage; authentication that addresses the verification of a user's identity and eligibility to data access; and confidentiality that requires data use is confined to authorized purposes by authorized people" (p. 996). Thus, security includes all steps to make the storage of personal information secure.

Security and privacy have certain aspects in common. Especially the improper access dimension of Smith's construct is related to security to the extent that a person might be able to get access to personal information. These cases include the well-known examples of security breaches such as hacker attacks or data theft by unauthorized persons. Thus, companies cannot protect the individual's privacy without security.

According to Ackerman (2004), security is a necessary but not sufficient precondition for the protection of an individual's privacy. Culnan and Williams (2009) as well as Solove (2006) also define security as being one part of privacy.

However, one distinctive feature is the ethical dimension. Even when the company has made every effort to ensure security, privacy can be still threatened by moral failings such as the unauthorized secondary use of information.

Culnan and Williams (2009) identify vulnerability and avoiding harm as the two parts of morality which are important in the relationship between companies and their customers. Vulnerability exists due to the asymmetrical distribution of information and control. The company has the power to decide how to deal with the information collected. Managers can treat customer information in accordance with ethical guidelines or they can harm the customers, for example, by unauthorized secondary use.

Foxman and Kilkoyne (1993) address ethical dimensions with regard to privacy and a company's marketing practices by showing corporate activities that

Table 5-1: Simulation of Privacy and Security Breach in the Laboratory Experiment

Privacy Breach	The bank is transmitting personal data to a cooperating insurance company without the client's permission.
Security Breach	The bank has lost customer data. A former bank employee has stolen a CD with personal information and is now offering it for sale.

potentially threaten consumers' privacy. They conclude that the relationship between firms and customers is seriously affected by privacy violations. Accordingly, the company should treat personal information in a way that is consistent with the customers' right to privacy. Straub and Collins (1990) believe that this right to privacy "can best be protected through self-regulating policies and procedures."

Our research builds on these observations on morality and ethics when differentiating between a privacy and security incident. The privacy violation in our study lies in the fact that the bank is transferring personal information to a cooperating insurance company without the client's permission. The security breach is a stolen CD with customer information which is now offered for sale (see Table 5-1). We assume that public opinion would differentiate between both cases. While the company does not fulfil its moral responsibilities in the case of the privacy breach, the security incident is caused by unauthorized access and thus a criminal activity.

5.3.3 Trust

Both privacy and security are important factors for building trust in a company. Trust is crucial in virtually all interpersonal relations and economic transactions (Hosmer 1995). The meaning of trust has been studied in various disciplines, such as psychology (Rotter 1971), sociology, (Granovetter 1985) and economics (Gefen 2000). This is why many definitions exist, often reflecting the perspectives from the different disciplines, but today most researchers see it as a multidimensional and context-dependent construct (Ganesan 1994; Rousseau et al. 1998).

Gefen et al. (2003) provide a detailed overview of previous conceptualizations of trust in the literature. Although definitions vary across disciplines, researchers from different disciplines agree upon some necessary conditions for trust. Trust becomes relevant if the situation involves uncertainty about the future outcomes, because the trustor does not have the complete control and must enter into risks, being dependent on the decisions of the trustee who can either act trustworthy or untrustworthy (Kee and Knox 1970). The relationship between trust and risk is a reciprocal one, "risk creates an opportunity for trust, which leads to risk taking" (Rousseau et al. 1998). There would be no need for trust if there was complete certainty about the behavior of the acting persons. The trustor will rely upon the trustee if he perceives three characteristics to be met (Bhattacherjee 2002; McKnight et al. 2002): ability (concerns about the competence of the trustee), integrity (concerns about the honesty and moral principles) and benevolence (concerns about the goodwill towards the trustor). The nature of trust depends on the degree of interdependence – another necessary condition – which means the reliance between trustor and trustee. Researchers across disciplines also see trust as a psychological condition, rather than a behavior or choice.

The necessary conditions for trust are reflected in the popular notion of Mayer et al. (1995) who define trust as "the willingness of a party to be vulnerable to the actions of another party based on the expectation that the other will perform a particular action important to the trustor, irrespective of the ability to monitor or confront that other party". In the context of this paper it is important to distinguish between general trust and initial trust. General trust develops over time based on experiences between the trusting party and the trustee. We focus on initial trust, which occurs "when parties first meet or interact" (McKnight et al. 1998). In this situation neither of the two parties has any kind of experiences by means of which the trustworthiness could be evaluated.

5.4 Research Model

For the purpose of our study, the following research model can be derived from previous academic work (see Figure 5-1).

We assume both a direct impact of privacy and security on the actual behavior and an indirect relationship between privacy, security, trust, and behavior. We will focus on the case from the financial industry and will examine the impact of privacy and security incidents on the investment behavior (i.e., purchase of a financial product offered by the bank).

Previous research shows that privacy and procedural fairness are important antecedents of trust. Consumers' trust in e-commerce companies, for example, is positively affected by the level of privacy protection and the attempts of the firm to ensure data security (Suh and Han 2003). Moreover, the perception of how the company is treating customer data also impacts this relationship (Liu et al. 2005). Gefen et al. (2003) found that the trust in an e-vendor increases when customers believe that the vendor does not gain any advantages from being untrustworthy. The authors also show that security mechanisms on a website are important antecedents of trust. Based on the results from an analysis of industries

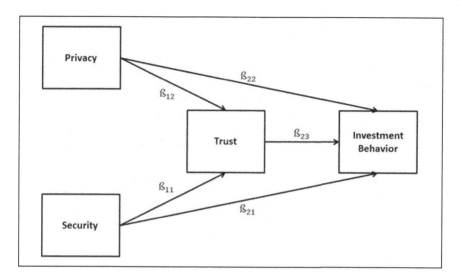

Figure 5-1:   Conceptual Framework

employing database marketing strategies, Milne and Boza (1999) infer that trust can be influenced by the likelihood that an organization is sharing information with third parties.

Hence, companies should behave in line with their own privacy policy, since consumers' expectations regarding what will be done with their data is built upon these organizational regulations (Culnan and Armstrong 1999). If a bank for example is transmitting customer information to a cooperating insurance company without the clients' permission and without mentioning it explicitly in their privacy disclosure, people might be displeased. As a result, one can expect that the trust in the company will suffer due to this privacy violation.

Hence, we hypothesize:

H1a: A security breach at a company has a negative impact on trust in the company.

H1b: A privacy violation by a company has a negative impact on trust in the company.

Only few studies examined the direct link between privacy, security, and purchase intentions. One of these studies was conducted by Eastlick et al. (2006) who found that consumers' privacy concerns can have a negative impact on the purchase intention towards an e-tailer. These findings are consistent with the results of studies in the field of direct marketing, suggesting that privacy concerns negatively influence purchase levels and direct marketing response (Milne and Boza 1999).

We will examine a case where from a rational point of view, people should behave equally, no matter whether a security or privacy problem exists or not, as the expected outcome does not change. However, empirical results suggest that the economic behavior can be influenced by feelings and emotions. For instance, people are more optimistic when they are in a good mood (Schwartz and Clore 1983) and risks can be judged differently, depending on the way the information is presented (Johnson and Tversky 1983). While standard finance theory posits that people act rationally, behavioral finance theory includes cognitive errors of human behavior (Statman 1999). For instance, there is evidence for the overreaction of stock markets following unexpected news events, as investors overweight recent information and underweight earlier data (De Bondt and Thaler 1985). It is therefore likely that privacy and security problems negatively impact the consumers' investment decision.

Collectively, these findings suggest:

H2a: A security breach at a company has a negative, direct impact on consumer behavior (here: investment behavior).

H2b: A privacy violation by a company has a negative, direct impact on consumer behavior (here: investment behavior).

Researchers focus more frequently on the impact of trust on behavioral intentions, particularly within business relationships. Many studies in e-commerce show that trust is a crucial determinant for the intention to buy a product. For instance, it was found that trust in the vendor significantly influences people's intention to purchase books on amazon.com (Gefen 2000). By studying the online shopping behavior of undergraduate students, Kim et al. (2008) show that consumers' trust influences not only the purchase intention, but also the actual purchase behavior. The authors invited students to visit at least two shopping websites and to search for products they were interested in. Before confirming the purchase, they were assigned to one questionnaire, either with questions about the website they were more likely to buy from, or with questions about the website they were less likely to buy from. Afterwards participants continued their purchase from the preferred website. The model created by McKnight and Chervany (2002) also posits that the customer is more likely to purchase from a company if the company's behavior seems to be honest and predictable. As people perceive their financial information as especially sensitive (Woodman et al. 1982), the role of trust could also be important for investment decisions.

Hence, we hypothesize:

> H3: Trust in a company positively impacts consumer behavior (here: amount of investment).

A great deal of studies and surveys show that there seem to be growing concern among consumers who fear that their personal information is not protected enough. According to a Gallup poll, 65 percent of Facebook users and 52 percent of Google users are worried about their privacy when using these internet applications (Morales, 2011). However, there is evidence that the actual behavior does not always reflect these general privacy concerns. The difference between intentions and behavior was described as the "privacy paradox" in the literature (Norberg et al. 2007; Smith et al. 2011). For instance, Spiekermann et al. (2001) compared the disclosing behavior of online shoppers with their previously stated privacy concerns. Surprisingly, participants have been willing to provide a great deal of private information (e.g., address), although reporting to be highly concerned about their personal data. Norberg et al. (2007) also show that people actually disclose far more personal information (e.g., financials, demographics) to a commercial enterprise than they intend to disclose. The dichotomy between stated intentions and actual behavior with regard to privacy suggests that people's trust in a company is more affected by a privacy breach than their behavior. We therefore assume that the intention-behavior gap persists after the occurrence of a privacy breach.

Hence, we hypothesize:

> H4: A privacy violation by a company has a stronger negative impact on trust than on actual consumer behavior.

5.5 Laboratory Experiment

5.5.1 Method

In contrast to previous research that used event study methodology for showing the reaction of the capital market (second-order effect), we conducted a laboratory experiment in order to focus on the direct consumer reaction (first-order effect) to privacy and security incidents. Although this is an artificial environment and one must be careful when generalizing findings, there are many advantages of experiments: researchers have the opportunity to effectively manipulate the independent variables and control for other influences so that causal relationships can be identified, which is an advantage compared to other methods including event studies. Another reason for the popularity of this method is the possibility of an inexpensive implementation and replication (Emory and Cooper 1991) that allows one to test the robustness of the findings.

The task for each subject was to decide about a financial investment. We used the investment decision as a cover story and did not reveal the real purpose of our study, namely the consumer reaction to different data protection violations. Cover stories have been successfully used in consumer research (e.g., Childers and Houston 1984; Gorn 1982). For instance, the cover story of Gorn (1982) comprised the selection of music for a pen commercial by the participating subjects. The actual purpose of the study was to show the relationship between the choice of the pen and the kind of music that was being played. Subjects were more likely to pick the color of the pen that was paired with liked rather than disliked music.

We applied a between-subject design, where all participants were randomly assigned to one of three groups. Every group received exactly the same information on the characteristics of the bank and the investment conditions. The information only differed with regard to a small detail about the privacy or security incident in the recent past of the bank. The participants in the control group were not confronted with any privacy or security breach. In the first treatment group we added the following sentence to the general description of the bank "The bank has recently been caught transmitting personal data to a cooperating insurance company without the client's permission." This clearly describes a privacy violation. In the second treatment group we added the sentence: "The bank has lost customer data. A former bank employee has stolen a CD with personal information and is now offering it for sale." This describes the security breach. This additional treatment information was presented very shortly at the end of the bank description.

Participants had to indicate the amount of money they were willing to invest into a financial product given the investment plan offered by the bank. To create an economic decision situation that reflects this decision, we modified the so-called "investment game", first introduced by Berg et al. (1995). This experimental method allows the measurement of trust in another person by the following procedure: one person, the trustor, receives 10 US dollars which can be invested into a geographically separated person (the trustee) who is unknown to the trustor. As soon as the trustee receives the money, the invested sum is tripled. The trustee now can decide how much money s/he is willing to send back to the trustor and how much s/he will keep for her/his own. It is certainly rational for the trustee to keep all money as s/he does not know the trustor and this is a one-shot-game. The trustor can of course anticipate this behavior and should, from an economic point of view, not invest any money in the trustee. However, several experimental studies show that money is invested and people tend to trust even unknown persons (Bolle 1998; Forsythe et al. 1994). Thus, in this game, trust can lead to monetary gains.

In our case, we conducted a slightly adapted investment game. The trustee is not another person but the trustor has to decide how much money s/he is willing to invest into a financial product offered by a fictional bank. In the experimental instructions we provided information on the fictitious bank which was similar to those of real banks, as well as the conditions under which they could invest their own real money: in all groups the investment horizon was 10 years in which the performance of the invested capital was 7 percent per year, given a default rate of 10 percent. The subjects received EUR 10 in cash and were offered the possibility to invest this money. They could invest up to EUR 10 and received their interest-paying money back with a probability of 90 percent (=1-default rate) after the experiment which lasted about 15 minutes. However, there was no obligation to invest a share so that participants could also keep all the money and leave immediately. In this case, there was no chance to generate more than EUR 10 but also no risk to lose the money due to the default of the bank (i.e., an unlucky die roll).

In order to illustrate the rules, subjects received the following numerical example: "Assume that you invest EUR 5, then you keep the other EUR 5 in all cases. The invested capital is virtually doubled given that there is no default, for which the probability is 10 percent. Thus, the complete amount paid out is EUR 15 at the end of the experiment if there is no default, otherwise it is EUR 5." This example clarifies that there is an element of risk since the repayment of the invested money depends on the default of the bank.

Based on the roll of a die, every 10th participant did not get his investment back. The probability of a default was totally independent of privacy or security incidents. Differences among the different experimental groups in terms of trust and behavior are therefore irrationally caused by the different levels of privacy and security concerns. The uncertainty about future returns due to the possible default leads to a trust game between the trustor (= participant) and the trustee (= fictional bank). If participants place more trust in the bank, they are likely to invest a higher proportion of their capital.

In order to determine subjects' trust in the bank, we used a 7 item Likert scale (1 = strongly disagree, 7 = strongly agree, see appendix). This scale aims to measure trust as beliefs about the other party's honesty, dependability, reliability, and trustworthiness (Pavlou and Gefen 2004). We also control for demographic information since family status and gender have been previously shown to exert an influence on trust (Buchan et al. 2008; Gilbert and Tang 1998) as well as on the investment behavior (Barber 2001; Cohn 1975).

5.5.2   Results

We recruited 118 undergraduate students on the university campus in order to participate in an investment experiment (cover story). We conducted the experiment in dedicated PC pools.

Descriptive statistics

The average age of the students is 24 years, 88 out of the 118 participants are aged between 21 and 26. It should also be noted that the average income is rather low. The majority has a monthly income of EUR 900 or less. Only 2 participants are married, 61 participants live alone and 55 participants live in a relationship. These numbers are not very surprising due to the University background. On average, subjects invest EUR 6.07 into the fictional product of the bank.

While subjects in the control group, who were not confronted with any privacy or security incident, invest on average EUR 7.41 of their capital, this amount is reduced by EUR 1 (-16 percent) in case of a privacy violation and by EUR 3 (-39 percent) when a security breach leads to data theft. These numbers suggest that security breaches have a higher economic impact than privacy breaches. The following analysis will clarify the influence of both incidents on trust and the investment amount.

Model

With the following set of equations we tested our hypotheses.

$$\text{Trust}_i = \alpha_1 + \text{\ss}_{11} * \text{Security}_i + \text{\ss}_{12} * \text{Privacy}_i + \text{\ss}_{13} * \text{FamilyStatus}_i + \text{\ss}_{14} * \text{Gender}_i + e_i \tag{5.1}$$

$$\text{IA}_i = \alpha_2 + \text{\ss}_{21} * \text{Security}_i + \text{\ss}_{22} * \text{Privacy}_i + \text{\ss}_{23} * \text{Trust}_i + \text{\ss}_{24} * \text{FamilyStatus}_i + \text{\ss}_{25} * \text{Gender}_i + e_i \tag{5.2}$$

where Security_i is a dummy variable indicating whether a security breach occurred (1 = security breach, 0 otherwise); Privacy_i is a dummy variable indicating whether a privacy breach occurred (1 = privacy breach, 0 otherwise); Trust_i is the amount of trust of person i into the bank. IA_i is the investment amount that a person i is willing to invest into the bank. Subjects also provided information on family status (single=1, in a relationship=2, married=3) as well as gender (1=female; 2=male).

We used seemingly unrelated regression analyses (SURE) as well as OLS in order to estimate the sets of equations (1) and (2). SURE method was introduced by Zellner (1962) for estimating regressions where disturbances correlate. In our case, trust as measured by the 7 item Likert scale is the dependent variable in equation 5.1 and is used as an independent variable in equation 5.2.

Results

The assumptions of the model are fulfilled. Problems with multicollinearity do not exist since all VIFs are below 4 (mean VIF Regression 1 on trust = 1.18; Regression 2 on investment amount = 1.24). Breusch-Pagan test reveals that there is no heteroskedasticity so that we do not have to use robust standard errors.

As Table 5-2 illustrates, both privacy and security incidents negatively affect the amount of trust in the bank, supporting H1a (p<.01) and H1b (p<.01). This is not very surprising and supports previous findings. Our study allows however assessing the impact of privacy and security incidents with respect to behavior and in monetary terms. First, we find that trust has a positive impact on behavior which support hypothesis H3 (p<.01). We further observe that a security breach negatively influences the willingness to invest, supporting hypothesis H2a (p<.01). This result is interesting as it indicates that there is some additional

Table 5-2: Impact of Security and Privacy Incidents on Trust and Investment

	Coeffi-cient	Std. Error	t-Value	Coeffi-cient	Std. Error	t-Value
Dependent variable: Trust in Bank						
	SURE			**OLS**		
α_1 (Constant)	5.07 ***	0.60	8.44	5.07 ***	0.61	8.26
β_{11} (Security breach)	-1.04 ***	0.29	-3.56	-1.04 ***	0.30	-3.48
β_{12} (Privacy breach)	-1.17 ***	0.29	-3.97	-1.17 ***	0.30	-3.88
β_{13} (Family status)	-0.08	0.23	-0.36	-0.08	0.23	-0.35
β_{14} (Gender)	-0.11	0.25	-0.44	-0.11	0.26	-0.43
*** p < .01; Observations = 118; R^2 = .14						
Dependent variable: Investment amount						
	SURE			**OLS**		
α_2 (Constant)	1.64	1.77	0.93	1.64	1.81	0.91
β_{21} (Security breach)	-1.88 ***	0.72	-2.63	-1.88 **	0.74	-2.56
β_{22} (Privacy breach)	-0.08	0.73	-0.11	-0.08	0.75	-0.11
β_{23} (Trust in bank)	0.95 ***	0.21	4.46	0.95 ***	0.22	4.34
β_{24} (Family status)	0.67	0.53	1.26	0.67	0.55	1.23
β_{25} (Gender)	0.14	0.58	0.24	0.14	0.60	0.23
*** p < .01; ** p < .05; Observations = 118; R^2 = .25						

latent influence of security breaches above and beyond the indirect influence through trust. Security breaches thus hurt the relationship to the bank by lowering trust and above and beyond this impact there is some latent influence that additionally lowers the willingness make business with this bank.

If we look at the impact of privacy violations on the investment amount, we do not observe a significant effect (p>.1). There is no direct influence of privacy violations on behavior beside the indirect effect through trust. We therefore have to reject H2b but we find support for H4 that privacy significantly exerts a stronger negative impact on trust (-1.17) than on the investment amount. This result empirically supports the privacy paradox, which means that privacy influences intentions and behavior differently. However, one has to remember that trust influences behavior (hypothesis H3) and privacy issues influences trust (hypothesis H1b) and therefore an indirect influence still exists.

5.5.3   Robustness Check

In order to test whether our student sample is representative, we conducted a survey among the total population in Germany. Overall, 216 individuals took part in the nationwide survey. Our goal was to compare the privacy concerns as well as knowledge and experience of the students with the total population. We used the four dimensions of Smith's (1996) instrument: Errors, unauthorized secondary use, collection and improper access. These dimensions contain privacy and security statements and subjects specify their agreement (e.g., "Computer databases that contain personal information should be protected from unauthorized access - no matter how much it costs") on a 7 point Likert scale. While the student sample has an average score of 5.597, privacy/security concerns of the total population have an average level of 5.573. These differences in concerns are statistically not significant (t-test, p>.10). Thus, results reveal that our students have the same level of privacy concerns as the total population

We also collected information about the knowledge by asking whether subjects are aware of privacy and security risks and whether they have been a victim of a breach in the past (i.e. data theft). Results reveal a large amount of knowledge regarding privacy and security. Again, t-test (p>.10) revealed no significant differences between both groups so that we can assume that our student sample is representative for studying the effect of privacy and security breaches on the investment behavior. While privacy concerns differ across countries (e.g., Dinev et al. 2006), they seem to be stable within one society. We therefore expect the same investment behavior of the total population, which is however subject to future research projects and cannot be finally clarified in this study.

5.6 Discussion

5.6.1   Summary

To the best of our knowledge, this is the first study quantifying the impact of privacy and security incidents by performing a laboratory experiment. While the general, indirect link between privacy, security, trust, and behavioral intention has been extensively studied in literature, the direct impact of privacy and security breaches has received less attention so far. Our results clearly reveal a first-order effect, that is, a direct consumer reaction to privacy and security incidents.

A surprising result at first sight is the stronger impact of the security breach on the investment amount. One explanation could be that people perceive their financial information as especially sensitive (Woodman et al. 1982) and therefore fear that criminals can get access to their data. With regard to the serious monetary consequences that can result from abuse of account passwords or credit card numbers, a bank customer might be primarily interested in the security of his personal data. Another reason might be that people already assume secondary use of information to some extent, since many cases of privacy violations have been reported in the press.

Thus, meanwhile, the transfer of personal information to another company might be perceived as unpleasant, but also as a conventional business practice that clearly lowers trust in the long-term but does not affect the real investment decision in the same way. For their investments people seem to be primarily interested in the competence of the bank, i.e. the ability to manage the money and to provide secure data systems. The experiment shows that privacy issues influence behavior only indirectly through trust while security issues influence behavior directly above the indirect influence through trust.

Our study therefore contributes to a better understanding of the privacy paradox which has been previously shown in the literature (see section 5.4). In contrast to previous research, we study consumer behavior after a privacy breach actually occurred. So far, intentions and behavior have only been compared in the absence of any privacy or security incident. Although privacy is of prime importance for building trust, we find that following a privacy breach, people still ignore their concerns when it comes to the actual investment decision. We can therefore conclude that a privacy breach lowers trust in the company but does not exert a direct influence on consumers' willingness to buy products from the affected company.

The consequences of these results for overall welfare can be illustrated by looking at the allocation of financial assets. In 2009, every German citizen held about EUR 16,628 of his/her capital in securities . We can easily assume that the bank, that played the role model for our fictional bank, has a total of 15 million clients and around 400,000 new customers per year. These customers own securities worth approximately EUR 6.65 billion. A reduction of the investments by 39 percent (16 percent) would decrease the invested capital by EUR 2.59 (1.06) billion. If we assume an interest rate of 7 percent, this mistrust would cause a decrease of welfare by about EUR 182 million.

Recent data protection incidents show that companies around the world face enormous threats in this area. Every organization can easily become a target of cyber-attacks and data thefts. Hence, investment in security is required and this study introduces one method that allows assessing the expected monetary losses due to criminal activities which can be used to conduct costs-benefit analysis.

5.6.2   Limitations and Future Research

One limitation of our study is that the experiment was conducted in Germany, where data privacy is of a rather high value for the citizens compared to other countries (Singh and Hill 2003). This is also reflected by the stringent German laws, and one would expect that German consumers have high expectations with regard to data protection and get easily upset in case of privacy violations. This could lead to an overestimation of the impact of privacy violations.

There are already signs in the literature indicating differences in privacy concerns across societies. Bellman et al. (2004) found cultural values as an explanation for different levels of privacy concerns in 38 countries. Cho et al. (2009) showed that internet users in Asia have less privacy concerns compared to western countries. According to Dinev et al. (2006), Italians have less privacy concerns than US citizens.

Cultural values also influence legislation. Milberg et al. (2000) found that the level of privacy concern exerts a positive influence on regulatory preferences for strong laws as well as government involvement. The authors conclude that "a universal regulatory approach to information privacy seems unlikely and would ignore cultural and societal differences." It is therefore possible that trust in the company is affected differently across countries, depending on laws and privacy concerns. Cross-cultural differences could be tested in future experimental studies.

Another avenue for future research is a further examination of the trust relationship between the company and the consumer. We focused on initial trust in this study as subjects in our sample had no prior experience with the bank and were only informed about the company by our instruction. In a long-term relationship, customers have multiple interactions and can develop trust based on their experiences with regard to the bank's service, reliability and overall behavior. Thus, future research can take these circumstances into account and focus on the reactions of existing investors to privacy and security problems.

In particular, there might be positive effects of security breaches on trust. Given that the bank makes great efforts to improve security measures, customers might perceive transactions with this bank as extremely secure. In our experiment, we informed subjects that the security breach occurred recently and that the CD is now circulating in the market place. Thus, the bank had probably not enough time to revise their security strategy. However, positive effects on trust might still be possible and can be specifically investigated in future research projects.

We took a bank as an example to quantify the effects of privacy and security incidents. It would be interesting to compare the results with other industries, since customers usually express grave concerns about their bank data.

A further limitation, but similar to the original investment game setting of Berg et al. (1995), is the student sample. The impact of privacy and security breaches on the investment behavior might not be representative for the overall society. In our case, however, this limitation should not be severe as we are mainly interested in differences and not in absolute values. Moreover, the subjects in the sample are very likely to be important new customers and new investors in the near future.

Moreover, due to the results of our robustness check, we assume that our student sample is representative for the total population. We find evidence that privacy concerns do not differ across the society and we also observe the same privacy knowledge and experience. One can therefore expect the same investment behavior of the entire population when it comes to privacy and security incidents.

In sum, we are confident that our laboratory experiment is a good proxy for real behavior. The experiment allows a high level of control, which is very hard to realize in a field experiment or event studies. Furthermore, from a practical point of view, it appears rather unlikely to find a bank that is willing to simulate privacy or security breaches in order to conduct a field experiment.

We conclude that privacy and security breaches harm both the company as well as overall welfare. Further research in this area can help organizations to better understand the importance of data protection and the impact of security incidents and to take appropriate measures regarding the clients' protection with regards to privacy and security threats.

5.7 Appendix

Information on the bank

- Founded in 1870
- Total assets: EUR 844.1 billion
- Number of employees: 62,000
- Second largest German bank
- 15 million private and business clients

Investment conditions

- Duration: 10 years
- Starting capital: EUR 10
- Rate of return: 7% p.a.
- Default risk: 10%, that is the capital is repaid at the end of the term with a probability of 90%. You receive the money immediately following the experiment.
- Payout factor: 0.5

Example: If you invest EUR 5, then you keep the other EUR 5 of the starting capital. In addition, the invested EUR 5 are doubled, given that there is no default. The complete payoff is EUR 15 (5+2x5), if there is no default. The payment is multiplied with the payout factor of 0.5.

Measurement of Trust on a 7 item Likert scale (Pavlou and Gefen 2004)

1. According to the information provided the described bank seems to be dependable.
2. According to the information provided the described bank is reliable and a serious trading partner.
3. The described bank is honest with regard to its statements.
4. The described bank is trustworthy in general.

(1=strongly disagree, 7=strongly agree, 4=neither agree nor disagree)

6 Literature

Ackerman M (2004) Privacy in Pervasive Environments: Next Generation Labeling Protocols. Personal and Ubiquitous Computing 8(6):430-439

Acquisti A, Friedman A, Telang R (2006) Is There a Cost to Privacy Breaches? An Event Study. In: Proceedings of the Twenty Seventh International Conference on Information Systems, Milwaukee

Akerlof GA (1970) The Market for 'Lemons': Quality Uncertainty and the Market Mechanism. Quarterly Journal of Economics 84(3):353–374

Ali K, Pazzani M (1992) Reducing the Small Disjuncts Problem by Learning Probabilistic Concept Descriptions. In Petsche T, Judd S, Hanson S (eds.): Computational Learning Theory and Natural Learning Systems, Vol. 3. MIT Press, Cambridge

Ammon U (2009) Delphi-Befragung - Handbuch Methoden der Organisationsforschung. VS Verlag für Sozialwissenschaften, Wiesbaden

Anderson JR (1981) Cognitive Skills and Their Acquisition. Erlbaum, New Jersey

Andoh-Baidoo FK, Amoako-Gyampah K, Osei-Bryson KM (2010) How Internet Security Breaches Harm Market Value. IEEE Security and Privacy 8(1):36-42

Antweiler W, Frank MZ (2004) Is All That Talk Just Noise? The Information Content of Internet Stock Message Boards. Journal of Finance 59(3):1259-1294

Aral S, Brynjolfsson E, Van Alstyne M (2008) Sharing Mental Models: Antecedents and Consequences of Mutual Knowledge in Teams. Working Paper

Arch E (1993) Risk-taking: A Motivational Basis for Sex Differences. Psychological Reports 73(3):6-11

Ariel RA (1987) A Monthly Effect in Stock Returns. Journal of Financial Economics 18(1):161-174

Ariel RA (1990) High Stock Returns before Holidays: Existence and Evidence on Possible Causes. Journal of Finance 45(5):1611-1626

Armstrong JS (1980) The Seer-Sucker Theory: The Value of Experts in Forecasting. Technology Review 83:16-24

Avery C, Chevalier J, Zeckhauser R (2009) The 'CAPS' Prediction System and Stock Market Returns. Working Paper, Harvard Kennedy School

Ba S, Pavlou P (2002) Evidence of the Effect of Trust Building Technology in Electronic Markets: Price Premiums and Buyer Behavior. MIS Quarterly 26(3):243–268

Bajtelsmit VL, VanDerhei JA (1997) Risk Aversion and Pension Investment Choices. In: Gordon MS, Mitchell OS, Twinney MM (eds.) Positioning Pensions for the Twenty-First Century. University of Pennsylvania Press, Philadelphia, pp. 91-103

Baker M, Stein J (2004) Market Liquidity as a Sentiment Indicator. Journal of Financial Markets 7(3):271–99

Baker M, Wurgler J (2007) Investor Sentiment in the Stock Market. Journal of Economic Perspectives 21(2):129–151

Bakos Y (1998) The Emerging Role of Electronic Marketplaces on the Internet. Communications of the ACM 41(8):35-42

Bakshy E, Hofman JM, Mason WA, Watts DJ (2011) Everyone's an Influencer: Quantifying Influence on Twitter. Proceedings of the Fourth ACM International Conference on Web Search and Data

Banerjee AV (1992) A Simple Model of Herd Behavior. Quarterly Journal of Economics 107(3):797-817

Bantel KA, Jackson SE (1989) Top Management and Innovations in Banking: Does the Composition of the Top Team Make a Difference? Strategic Management Journal 10(S1):107-124

Bapna R, Goes P, Gupta A, Jin Y (2004) User Heterogeneity and its Impact on Electronic Auction Market Design: An Empirical Exploration. MIS Quarterly 28(1):21-43

Barber B, Odean T (2001) Boys Will Be Boys: Gender, Overconfidence, and Common Stock Investment. Quarterly Journal of Economics 116(1):261-292

Barber BM, Lehavy R, McNichols M, Trueman B (2006) Buys, Holds, and Sells: The Distribution of Investment Banks' Stock Ratings and the Implications for the Profitability of Analysts' Recommendations. Journal of Accounting and Economics 41(1-2):87-117

Belanger F, Hiller JS, Smith WJ (2002) Trustworthiness in Electronic Commerce: The Role of Privacy, Security, and Site Attributes. Journal of Strategic Information Systems 11(3-4):245-270

Bell NJ, Schoenrock CJ, O'Neal KK (2000) Self-Monitoring and the Propensity for Risk. European Journal of Personality 14(2):107-119

Bellman S, Johnson EJ, Kobrin SJ, Lohse GL (2004) International Differences in Information Privacy Concerns: A Global Survey of Consumers. Information Society 20(5):313-324

Benedict R (1934) Patterns of Culture. Houghton Mifflin, Boston

Bennouri M, Gimpel H, Robert J (2011) Measuring the Impact of Information Aggregation Mechanisms: An Experimental Investigation. Journal of Economic Behavior & Organization 78(3):302-318

Berendt B, Günther O, Spiekermann S (2005) Privacy in E-Commerce: Stated Preferences vs. Actual Behavior. Communications of the ACM 48(4):101-106

Berg J, Dickhaut J, Mccabe K (1995) Trust, Reciprocity, and Social History. Games and Economic Behavior 10(1):122-142

Berg J, Forsythe R, Rietz T (1997) What Makes Markets Predict Well? Evidence from the Iowa Electronic Markets. In: Albers W, Güth W, Hammerstein P, Moldovanu B, Van Damme E (eds.) Understanding Strategic Interaction: Essays in Honor of Reinhard Selten. Springer-Verlag, New York, pp. 444-463

Bhattacherjee A (2002) Individual Trust in Online Firms: Scale Development and Initial Test. Journal of Management Information Systems 19(1):211-241

Biemann C, Bordag S, Heyer G, Quasthoff U, Wolff C (2004) Language Independent Methods for Compiling Monolingual Lexical Data. In: Proceedings of CICLing, LNCS 2945:217-228

Bifet A, Frank E (2010) Sentiment Knowledge Discovery in Twitter Streaming Data. Discovery Science 6332:1-15

Bikhchandani S, Hirshleifer D, Welch I (1992) A Theory of Fads, Fashion, Custom, and Cultural Change as Informational Cascades. Journal of Political Economy 100(5):992-1026

Bilton N, Stelter B (2011) Sony Says PlayStation Hacker Got Personal Data. Last retrieved 2013-09-23, http://www.nytimes.com/2011/04/27/technology/27playstation.html?_r=0

Black F (1986) Noise. Journal of Finance 41(3):529-543

Bogle JC (2001) John Bogle on Investing. McGraw-Hill, New York

Bogle JC (2005) The Mutual Fund Industry Sixty Years Later: For Better or Worse? Financial Analysts Journal 61(1):15-24

Bolle F (1998) Rewarding Trust: An Experimental Study. Theory and Decision 45(1):83-98

Bollen J, Mao H, Zeng X (2010) Twitter Mood Predicts the Stock Market. Journal of Computational Science 2(1):1-8

Bono JE, Ilies R (2006) Charisma, Positive Emotions and Mood Contagion. The Leadership Quarterly 17:317-334

Boucher J, Osgood CE (1969) The Pollyanna Hypothesis. Journal of Verbal Learning and Verbal Behavior 8(1):1-8

Boucher JD (1979) Culture and Emotion. In Marsella AJ, Tharp R, Ciborowski T (eds.): Perspectives on Cross-Cultural Psychology, pp. 159-178, Academic Press, New York

Boyd D, Golder S, Lotan G (2010) Tweet, Tweet, Retweet: Conversational Aspects of Retweeting on Twitter. 43rd Hawaii International Conference on System Sciences (HICSS)

Brown R, Gilman A (1970) The Pronouns of Power and Solidarity. In Sebeok T (ed.): Style in Language, pp. 253-276. MIT press, Cambridge

Brown GW, Cliff MT (2005) Investor Sentiment and Asset Valuation. Journal of Business 78(2):405–40

Brown P, Keim DB, Kleidon AW, Marsh TA (1983) Stock Return Seasonalities and the Tax-Loss Selling Hypothesis: Analysis of the Arguments and Australian Evidence. Journal of Financial Economics 12(1):105-12

Brynjolfsson E, Smith MD (2000) Frictionless Commerce? A Comparison of Internet and Conventional Retailers. Management Science 46(4):563-585

Buchan NR, Croson RTA, Solnick S (2008) Trust and Gender: An Examination of Behavior and Beliefs in the Investment Game. Journal of Economic Behavior & Organization 68:466-476

Byrnes J, Miller DC, Schafer WD (1999) Gender Differences in Risk Taking: A Meta-Analysis. Psychological Bulletin 125(3):367-383

Caillaud B, Jullien B (2001) Competing Cybermediaries. European Economic Review 45(4-6):797-808

Campbell JY, Lo AW, MacKinlay AC (1997) The Econometrics of Financial Markets. Princeton University Press, Princeton

Cao HH, Coval JD, Hirshleifer D (2002) Sidelined Investors, Trade-Generated News, and Security Returns. Review of Financial Studies 15(2):615-648

Carhart MM (1997) On Persistence in Mutual Fund Performance. Journal of Finance 52(1):57-82

Carton S, Jouvent R, Bungener C, Widlöcher D (1992) Sensation Seeking and Depressive Mood. Personality and Individual Differences 13(7):843-849

Cavusoglu H, Mishra B, Raghunathan S (2004) The Effect of Internet Security Breach Announcements on Market Value: Capital Market Reactions for Breached Firms and Internet Security Developers. International Journal of Electronic Commerce 9(1):69-104

Cha M, Haddadi H, Benevenuto F, Gummadi KP (2010) Measuring User Influence in Twitter: The Million Follower Fallacy. Proceedings of the Fourth International AAAI Conference on Weblogs and Social Media

Chan K, Hameed A, Tong W (2000) Profitability of Momentum Strategies in the International Equity Markets. Journal of Financial and Quantitative Analysis 35(2):153-172

Chang S-C, Chen S-S, Chou RK, Lin Y-H (2008) Weather and Intraday Patterns in Stock Returns and Trading Activity. Journal of Banking & Finance 32:1754-176

Chang S-C, Chen S-S, Chou RK, Lin YH (2012) Local Sports Sentiment and Returns of Locally Headquartered Stocks: A Firm-Level Analysis. Journal of Empirical Finance 19(3):309-318

Chen G, Firth M, Rui OM (2001) The Dynamic Relation Between Stock Returns, Trading Volume, and Volatility. Financial Review 36(3):153-174

Cheng TC, Lam D, Yeung A (2006) Adoption of Internet Banking: An Empirical Study in Hong Kong. Decision Support Systems 42(3):1558-1572

Chesbrough H, Crowther AK (2006) Beyond High Tech: Early Adopters of Open Innovation in Other Industries. R&D Management 36(3):229-236

Chevalier JA, Mayzlin D (2006) The Effect of Word of Mouth on Sales: Online Book Reviews. Journal of Marketing Research 43(3):345-354

Childers TL, Houston MJ (1984) Conditions for a Picture-Superiority Effect on Consumer Memory. Journal of Consumer Research 11(2):643-654

Chintagunta PK, Gopinath S, Venkataraman S (2010) The Effects of Online User Reviews on Movie Box Office Performance: Accounting for Sequential Rollout and Aggregation Across Local Markets. Marketing Science 29(5):944-957

Cho H, Rivera-Sánchez M, Lim SS (2009) A Multinational Study on Online Privacy: Global Concerns and Local Responses. New Media & Society 11(3):395-416

Chordia T, Swaminathan B (2000) Trading Volume and Cross-Autocorrelations in Stock Returns. Journal of Finance 55(2):915-935

Chou KL, Lee T, Ho AH (2007) Does Mood State Change Risk Taking Tendency in Older Adults? Psychology and Aging 22(2):310

Choudhury V, Hartzel KS, Konsynski BR (1998) Uses and Consequences of Electronic Markets: An Empirical Investigation in the Aircraft Parts Industry. MIS Quarterly 22(4):471-507

Cocozza JJ, Steadman HJ (1978) Prediction in Psychiatry: An Example of Misplaced Confidence in Experts. Social Problems 25(3):265-276

Cohn RA, Lewellen WG, Lease RC, Schlarbaum GG (1975) Individual Investor Risk Aversion and Investment Portfolio Composition. Journal of Finance 30(2):605-620

Conover MD, Gonçalves B, Ratkiewicz J, Flammini A, Menczer F (2011) Predicting the political Alignment of Twitter Users. Proceedings of the International Conference on Social Computing

Coval JD, Moskowitz TJ (1999) Home Bias at Home: Local Equity Preference in Domestic Portfolios. The Journal of Finance 54(6):2045-2073

Coviello L, Sohn Y, Kramer AD, Marlow C, Franceschetti M, Christakis NA, Fowler JH (2014) Detecting Emotional Contagion in Massive Social Networks. PloS one 9(3):e90315

Culnan MJ (1993) How Did They Get My Name? An Exploratory Investigation of Consumer Attitudes Toward Secondary Information Use. MIS Quarterly 17(3):341-364

Culnan MJ, Armstrong PK (1999) Information Privacy Concerns, Procedural Fairness, and Impersonal Trust: An Empirical Investigation. Organization Science 10(1):104-115

Culnan MJ, Williams CC (2009) How Ethics Can Enhance Organizational Privacy: Lessons from the Choice Point and TJX Data Breaches. MIS Quarterly 33(4):673-687

D'Onfro J (2013) Twitter Admits 5% of its 'Users' Are Fake. Last retrieved 2014-06-22, http://www.businessinsider.com/5-of-twitter-monthly-active-users-are-fake-2013-10

Dalbert C (1992) Subjektives Wohlbefinden junger Erwachsener: Theoretische und empirische Analysen der Struktur und Stabilität. Zeitschrift für Differentielle und Diagnostische Psychologie 13:207-220

Dalkey N, Helmer O (1963) An Experimental Application of the Delphi Method to the Use of Experts. Management Science 9(3):458-467

Daniel K, Hirshleifer D, Teoh SH (2002) Investor Psychology in Capital Markets: Evidence and Policy Implications. Journal of Monetary Economics 49:139-209

Daniel K, Grinblatt M, Titman S, Wermers R (1997) Measuring Mutual Fund Performance With Characteristic-Based Benchmarks. Journal of Finance 52(3):1035-1058

Das SR, Sisk J (2003) Financial Communities. Santa Clara University, Working Paper.

Das SR, Chen MY (2007) Yahoo! for Amazon: Sentiment Extraction from Small Talk on the Web. Management Science 53(9):1375-1388

De Bondt WFM, Thaler R (1985) Does the Stock Market Overreact? Journal of Finance 40(3):793-805

De Long JB, Shleifer A, Summers LH, Waldmann RJ (1990) Positive Feedback Investment Strategies and Destabilizing Rational Speculation. Journal of Finance 45(2):379-395

Deaux K, Farris E (1977) Attributing Causes for One's Own Performance: The Effects of Sex, Norms, and Outcome. Journal of Research Personality 11(1):59-72

DeBondt WFM (1993) Betting on Trends: Intuitive Forecasts of Financial Risk and Return. International Journal of Forecasting 9(3):355-371

Dellarocas C (2003) The Digitization of Word of Mouth: Promise and Challenges of Online Feedback Mechanisms. Management Science 49(10):1407-1424

Dellarocas C, Zhang X, Awad NF (2007) Exploring the Value of Online Product Reviews in Forecasting Sales: The Case of Motion Pictures. Journal of Interactive Marketing 21(4):23-45

DeMarzo PM, Vayanos D, Zwiebel J (2003) Persuasion Bias, Social Influence, and Unidimensional Opinions. Quarterly Journal of Economics 118(3):909-968

Dhar V, Chang E (2009) Does Chatter Matter? The Impact of User-Generated Content on Music Sales. Journal of Interactive Marketing 23(4):300-307

Dichev ID, Janes TD (2003) Lunar Cycle Effects in Stock Returns. Journal of Private Equity 6(4):8-29

Diefenbach RE (1972) How Good is Institutional Brokerage Research? Financial Analysts Journal 28(1):54+56-60

Dimson E, Marsh P (1986) Event Study Methodologies and the Size Effect: The Case of UK Press Recommendations. Journal of Financial Economics 17(1):113-142

Dinev T, Bellotto M, Hart P, Russo V, Serra I, Colautti C (2006) Internet Users' Privacy Concerns and Beliefs About Government Surveillance: An Exploratory Study of Differences Between Italy and the United States. Journal of Global Information Management (14:4):57-93

Dommeyer CJ, Gross BL (2003) What Consumers Know and What They Do: An Investigation of Consumer Knowledge, Awareness, and Use of Protection Strategies. Journal of Interactive Marketing 17(2):34-51

Eastlick MA, Lotz SL, Warrington P (2006) Understanding Online B-to-C Relationships: An Integrated Model of Privacy Concerns, Trust, and Commitment. Journal of Business Research 59(8):877-886

eBay (2012) eBay Inc. Reports Strong Fourth Quarter and Full Year 2011 Results. Last retrieved 2012-03-27, http://investor.ebayinc.com/releasedetail.cfm?ReleaseID=640656

Edmans A, Garcia D, Norli Ø (2007) Sports Sentiment and Stock Returns. Journal of Finance 62(4):1967-1998

Eisenberg AE, Baron J, Seligman ME (1998) Individual Differences in Risk Aversion and Anxiety. Psychological Bulletin 87:245-251

Eisenmann T, Parker G, Van Alstyne MW (2006) Strategies for Two-Sided Markets. Harvard Business Review 84(10):92-101

Ellison G, Ellison SF (2005) Lessons about Markets from the Internet. Journal of Economic Perspectives 19(2):139-158

Elron E (1997) Top Management Teams within Multinational Corporations: Effects of cultural heterogeneity. Leadership Quart 8(4):393-412

eMarketer (2012) New Forecast: Emerging Markets Lead World in Social Networking Growth. Last retrieved 2014-09-19, http://www.emarketer.com/newsroom/index.php/forecast-emerging-markets-lead-world-social-networking-growth/#5qim7ZDrlU56sKSG.99

eMarketer (2013) Social Networking Reaches Nearly One in Four Around the World. Last retrieved 2014-09-18, http://www.emarketer.com/Article/Social-Networking-Reaches-Nearly-One-Four-Around-World/1009976

Evans D (2003) Some Empirical Aspects of Multi-Sided Platform Industries. Review of Network Economics 2(3):191-209

Fama EF (1970) Efficient Capital Markets: A Review of Theory and Empirical Work. Journal of Finance 25(2):383-417

Fama EF, French KR (1998) Value versus Growth: The International Evidence. Journal of Finance 53(6):1975-1999.

Fields MJ (1931) Stock Prices: A Problem in Verification. The Journal of Business of the University of Chicago 4(4):415-418

Fogel J, Nehmad E (2009) Internet Social Network Communities: Risk Taking, Trust, and Privacy Concerns. Computers in Human Behavior 25:153–160

Forgas JP (1995) Mood and Judgment: The Affect Infusion Model (AIM). Psychological Bulletin 117(1):39-66

Forsythe R, Rietz TA, Ross TW (1999) Wishes, Expectations and Actions: A survey on Price Formation in Election Stock Markets. Journal of Economic Behavior & Organization 39(1):83-110

Forsythe R, Horowitz JL, Savin NE, Sefton M (1994) Fairness in Simple Bargaining Experiments. Games and Economic Behavior 6(3):347-369

Foxman ER, Kilcoyne P (1993) Information Technology, Marketing Practice, and Consumer Privacy: Ethical Issues. Journal of Public Policy & Marketing 12(1):106-119

French KR, Schwert GW, Stambaugh RF (1987) Expected Stock Returns and Volatility. Journal of Financial Economics 19(1):3-29

French KR, Poterba JM (1991) Investor Diversification and International Equity markets. American Economic Review 81(2):222-226

Gallant AR, Rossi PE, Tauchen G (1992) Stock Prices and Volume. Review of Financial Studies 5(2):199-242

Galton F (1907) Vox Populi. Nature 75:450-451

Ganesan S (1994) Determinants of Long-Term Orientation in Buyer-Seller Relationships. Journal of Marketing 58(2):1-19

Gefen D (2000) E-Commerce: The Role of Familiarity and Trust. Omega 28(6):725-737

Gefen D, Karahanna E, Straub DW (2003) Trust and TAM in Online Shopping: An Integrated Model. MIS Quarterly 27(1):51-90

Gehm TL, Scherer KR (1988) Factors Determining the Dimensions of Subjective Emotional Space. In Scherer KR (ed.): Facets of Emotion: Recent Research, pp. 99-113. Lawrence Erlbaum Associates, Hillsdale

Ghosh R, Lerman K (2011) A Framework for Quantitative Analysis of Cascades on Networks. Proceedings of the Fourth ACM International Conference on Web Search and Data Mining

Gilbert E, Karahalios K (2010) Widespread Worry and the Stock Market. In: Proceedings of the Fourth International AAAI Conference on Weblogs and Social Media

Gilbert JA, Tang TLP (1998) An Examination of Organizational Trust Antecedents. Public Personnel Management 27(3):321-338

Giles J (2005) Internet Encyclopaedias Go Head to Head. Nature 438:900-901

Ginsberg J, Mohebbi MH, Patel RS, Brammer L, Smolinski ML, Brilliant L (2009) Detecting Influenza Epidemics Using Search Engine Query Data. Nature 457:1012-1015

Goh KY, Heng CS (2013) Social Media Brand Community and Consumer Behavior: Quantifying the Relative Impact of User- and Marketer-Generated Content. Information Systems Research 24(1): 88–107

Goodwin C (1991) Privacy: Recognition of a Consumer Right. Journal of Public Policy & Marketing 10(1):149-166

Gorn GJ (1982) The Effects of Music in Advertising on Choice Behavior: A Classical Conditioning Approach. Journal of Marketing 46:94-101.

Gottschlich J, Hinz O (2013) A Decision Support System for Stock Investment Recommendations Using Collective Wisdom. Decision Support Systems 59:52-62

Graham JR (1999) Herding among Investment Newsletters: Theory and Evidence. Journal of Finance 54(1):237-268

Grahl J, Hinz O, Rothlauf F (2014) What is the Value of a "Like"? – Experimental Evidence for the Influence of Popularity Signals on Shopping Behavior. Working Paper

Granovetter MS (1973) The Strength of Weak Ties. American Journal of Sociology 78(6):1360-1380

Granovetter M (1985) Economic Action and Social Structure: A Theory of Embeddedness. American Journal of Sociology 91(3):481-510

Greenaway KE, Chan YE (2005) Theoretical Explanations for Firms' Information Privacy Behavior. Journal of the Association for Information Systems 6(6):171-198

Grinblatt M, Keloharju M (2001) How Distance, Language, and Culture Influence Stockholdings and Trades. The Journal of Finance 56(3):1053-1073

Grinblatt M, Titman S, Wermers R (1995) Momentum Investment Strategies, Portfolio Performance, and Herding: A Study of Mutual Fund Behavior. American Economic Review 85(5):1088-1105

Groth JC, Lewellen WG, Schlarbaum GG, Lease RC (1979) An Analysis of Brokerage House Securities Recommendations. Financial Analysts Journal 35(1):32-40

Gu B, Konana P, Rajagopalan B, Chen HWM (2007) Competition among Virtual Communities and User Valuation: The Case of Investing-Related Communities. Information Systems Research 18(1):68-85

Guillory J, Spiegel J, Drislane M, Weiss B, Donner W, Hancock J (2011) Upset Now?: Emotion Contagion in Distributed Groups. Proceedings of the SIGCHI Conference on Human Factors in Computing Systems

Hancock JT, Gee K, Ciaccio K, Lin JMH (2008) I'm Sad You're Sad: Emotional Contagion in CMC. In Proceedings of the 2008 ACM Conference on Computer Supported Cooperative Work

Hatfield E, Cacioppo JT (1994) Emotional Contagion. University Press, Cambridge

Heckman JJ, Ichimura H, Todd PE (1997) Matching as an Econometric Evaluation Estimator: Evidence from Evaluating a Job Training Programme. Review of Economic Studies 64:605-654

Heimbach I, Hinz O (2012) How Smartphone Apps Can Help Predicting Music Sales. In: Proceedings of the 20th European Conference on Information Systems (ECIS), Barcelona, Spain

Hertwig R (2012) Tapping into the Wisdom of the Crowd - with Confidence. Science 336:303-304

Hidding GJ, Williams JR (2003) Are there First-Mover Advantages in B2B eCommerce Technologies? Paper presented 36th Annual Hawaii International Conference on System Sciences (HICSS'03) - Track 7, Hawaii.

Hill S, Ready-Campbell N (2011) Expert Stock Picker: The Wisdom of (Experts in) Crowds. International Journal of Electronic Commerce 15(3):73-102

Hinz O, Spann M (2008) The Impact of Information Diffusion on Bidding Behavior in Secret Reserve Price Auctions. Information Systems Research 19(3):351-368

Hinz RP, McCarthy DD, Turner JA (1997) Are Women Conservative Investors? Gender Differences in Participant Directed Pension Investments. In: Gordon MS, Mitchell OS, Twinney MM (eds.) Positioning Pensions for the Twenty-First Century. University of Pennsylvania Press, Philadelphia, pp. 91-103

Hinz O, Hann IH, Spann M (2011) Price Discrimination in E-Commerce? An Examination of Dynamic Pricing in Name-Your-Own-Price Markets. MIS Quarterly 35(1):81-98

Hinz O, Schulze C, Takac C (2013) New Product Adoption in Social Networks: Why Direction Matters. Journal of Business Research, forthcoming

Hinz O, Skiera B, Barrot C, Becker J (2011) Seeding Strategies for Viral Marketing: An Empirical Comparison. Journal of Marketing 75(6):55-71

Hirshleifer D, Shumway T (2003) Good Day Sunshine: Stock Returns and the Weather. Journal of Finance 58(3):1009-1032

Holte RC, Acker L, Porter BW (1989) Concept Learning and the Problem of Small Disjuncts. In Proceedings of the Eleventh International Joint Conference on Artificial Intelligence, Detroit, Michigan

Hong H, Stein JC (1999) A Unified Theory of Underreaction, Momentum Trading, and Overreaction in Asset Markets. Journal of Finance 54(6):2143-2184

Hong L, Page SE (2001) Problem Solving by Heterogeneous Agents. Journal of Economic Theory 97(1):123-163

Hong H, Kubik JD, Stein JC (2005) Thy Neighbor's Portfolio: Word-of-Mouth Effects in the Holdings and Trades of Money Managers. Journal of Finance 60(6):2801-2824

Hosmer LT (1995) Trust: The Connecting Link between Organizational Theory and Philosophical Ethics. Academy of Management Review 20(2):379-403

Howe J (2008) Crowdsourcing: Why the Power of the Crowd is Driving the Future of Business. Crown Business, New York

Huberman G (2001) Familiarity Breeds Investment. The Review of Financial Studies 14(3):659-680

Jaffe JF, Westerfield R, Ma C (1989) A Twist on the Monday Effect in Stock Prices: Evidence from the U.S. and Foreign Stock Markets. Journal of Banking & Finance 13(4-5):641-650

Jaffe J, Westerfield R (1985) The Week-End Effect in Common Stock Returns: The International Evidence. Journal of Finance, 40(2):433-454

Jegadeesh N, Titman S (1993) Returns to Buying Winners and Selling Losers: Implications for Stock Market Efficiency. Journal of Finance 48(1):65-91

Jehn KA, Northcraft GB, Neale MA (1999) Why Differences Make a Difference: A Field Study of Diversity, Conflict, and Performance in Workgroups. Administrative Science Quarterly 44(4):741-763

Jensen M (1968) The Performance of Mutual Funds in the Period 1945-1964. Journal of Finance 23(2):389-416

Jeppesen LB, Frederiksen L (2006) Why Do Users Contribute to Firm-Hosted User Communities? The Case of Computer-Controlled Music Instruments. Organization Science 17(1):45-63

Jianakoplos NA, Bernasek A (1998) Are Women More Risk Averse? Economic Inquiry 36(4):620-630

John LK, Acquisti A, Loewenstein G (2011) Strangers on a Plane: Context-Dependent Willingness to Divulge Sensitive Information. Journal of Consumer Research 37(5):858-873

Johnson EJ, Tversky A (1983) Affect, Generalization, and the Perception of Risk. Journal of Personality and Social Psychology 45(1):20-31

Johnston R, McNeal BF (1967) Statistical versus Clinical Prediction: Length of Neuropsychiatric Hospital Stay. Journal of Abnormal Psychology 72(4):335-340

Kalakota R, Whinston AB (1996) Frontiers of Electronic Commerce. Addison-Wesley, Reading

Kamstra MJ, Kramer LA, Levi MD (2000) Losing Sleep at the Market: The Daylight Saving Anomaly. American Economic Review 90(4):1005-1011

Kamstra MJ, Kramer LA, Levi MD (2003) Winter Blues: A SAD Stock Market Cycle. American Economic Review 93(1):324-343

Kaplan AM, Haenlein M (2010) Users of the World, Unite! The Challenges and Opportunities of Social Media. Business Horizons 59:59-68

Karabulut Y (2011) Can Facebook Predict Stock Market Activity? Working Paper, University of Frankfurt, Germany

Karpoff JM (1987) The Relation between Price Changes and Trading Volume: A Survey. Journal of Financial and Quantitative Analysis 22(1):109-126

Kee HW, Knox RE (1970) Conceptual and Methodological Considerations in the Study of Trust and Suspicion. Journal of Conflict Resolution 14(3):357-366

Kelly H (2013) Twitter Hacked; 250,000 Accounts Affected. Last retrieved 2013-09-23, http://edition.cnn.com/2013/02/01/tech/social-media/twitter-hacked/index.html

Kelly K, Low B, Tan HT, Tan SK (2012) Investors' Reliance on Analysts' Stock Recommendations and Mitigating Mechanisms for Potential Overreliance. Contemporary Accounting Research 29(3):991-1012

Kempe D, Kleinberg J, Tardos E (2003) Maximizing the Spread of Influence through a Social Network. Proceedings of the Ninth ACM SIGKDD International Conference on Knowledge Discovery and Data Mining

Kerin RA, Varadarajan PR, Peterson RA (1992) First-Mover Advantage: A Synthesis, Conceptual Framework, and Research Propositions. Journal of Marketing 56(4):33-52

Kilduff M, Angelmar R, Mehra A (2000) Top Management-Team Diversity and Firm Performance: Examining the Role of Cognitions. Organization Science 11(1):21-34

Kim DJ, Ferrin DL, Raghav Rao H (2008) A Trust-Based Consumer Decision-Making Model in Electronic Commerce: The Role of Trust, Perceived Risk, and their Antecedents. Decision Support Systems 44(2):544-564

Kittur A, Kraut RE (2008) Harnessing the Wisdom of Crowds in Wikipedia: Quality Through Coordination. In: Proceedings of the 2008 ACM conference on Computer supported cooperative work, pp. 37-46

Koriat A (2012) When are Two Heads Better than One and Why? Science 336:360-362

Kosner A (2013) Watch Out Facebook, With Google+ at #2 and YouTube at #3, Google, Inc. Could Catch Up. Last retrieved 2014-09-19, http://www.forbes.com/sites/anthonykosner/2013/01/26/watch-out-facebook-with-google-at-2-and-youtube-at-3-google-inc-could-catch-up/

Kramer AD (2012) The Spread of Emotion via Facebook. In Proceedings of the SIGCHI Conference on Human Factors in Computing Systems, pp. 767-770

Kramer AD, Guillory JE, Hancock JT (2014) Experimental Evidence of Massive-Scale Emotional Contagion through Social Networks. Proceedings of the National Academy of Sciences

Krämer W, Runde R (1997) Stocks and the Weather: An Exercise in Data Mining or yet another Capital Market Anomaly? Empirical Economics 22:637-641

Kwak H, Lee C, Park H, Moon S (2010) What Is Twitter, a Social Network or a News Media? Proceedings of the 19th International Conference on World Wide Web

Lakonishok J, Smidt S (1988) Are Seasonal Anomalies Real? A Ninety Year Perspective. Review of Financial Studies 1(4):403-425

Lakonishok J, Maberly E (1990) The Weekend Effect: Trading Patterns of Individual and Institutional Investors. Journal of Finance 45(1):231-243

Larkin JH, McDermott J, Simon DP, Simon HA (1980) Expert and Novice Performance in Solving Physics Problems. Science 208:1335-1342

Leetaru KH, Wang S, Cao G, Padmanabhan A, Shook E (2013) Mapping the Global Twitter Heartbeat: The Geography of Twitter. First Monday 18(5), last retrieved 2014-06-22 http://firstmonday.org/ojs/index.php/fm/article/view/4366/3654

Leimeister JM, Huber M, Bretschneider U, Krcmar H (2009) Leveraging Crowdsourcing: Activation-Supporting Components for IT-Based Ideas Competition. Journal of Management Information Systems 26(1):197-224

Lemmon M, Portniaguina E (2006) Consumer Confidence and Asset Prices: Some Empirical Evidence. Review of Financial Studies 19(4):1499–1529

Lerman K, Ghosh R, Surachawala T (2012) Social Contagion: An Empirical Study of Information Spread on Digg and Twitter Follower Graphs. Working Paper

Leskovec J, McGlohon M, Faloutsos C, Glance N, Hurst M (2007) Cascading Behavior in Large Blog Graphs. In: Proceedings of the 7th SIAM International Conference on Data Mining (SDM)

Levy BI, Ulman E (1967) Judging Psychopathology from Paintings. Journal of Abnormal Psychology 72(2):182-187

Levy T, Yagil J (2011) Air Pollution and Stock Returns in the US. Journal of Economic Psychology 32(3):374-383

Lewellen WG, Lease RC, Schlarbaum GG (1977) Patterns of Investment Strategy and Behavior among Individual Investors. Journal of Business 50(3):296-333

Lewis MP (2009) Ethnologue: Languages of the World. SIL International, Dallas

Liebowitz SJ, Margolis SE (1994) Network Externality: An Uncommon Tragedy. Journal of Economic Perspectives 8(2):133-150

Liu C, Marchewka JT, Lu J, Yu C (2005) Beyond Concern - A Privacy-Trust-Behavioral Intention Model of Electronic Commerce. Information & Management 42(1):289-304

Lo AW, Repin DV (2002) The Psychophysiology of Real-Time Financial Risk Processing. Journal of Cognitive Neuroscience 14(3):323-339

Loewenstein GF, Weber EU, Hsee CK, Welch N (2001) Risk as Feelings. Psychological Bulletin 127(2):267-286

Lorenz J, Rauhut H, Schweitzer F, Helbing D (2011) How Social Influence Can Undermine the Wisdom of Crowd Effect. In: Proceedings of the National Academy of Sciences of the United States of America 108(22):9020–9025

Luo X, Li H, Zhang J, Shim JP (2010) Examining Multi-Dimensional Trust and Multi-Faceted Risk in Initial Acceptance of Emerging Technologies: An Empirical Study of Mobile Banking Services. Decision Support Systems 49(2):222-234

MacKinlay AC (1997) Event Studies in Economics and Finance. Journal of Economic Literature 35(1):13-39

Malkiel BG (1995) Returns from Investing in Equity Mutual Funds 1971-1991. Journal of Finance 50(2):549-572

Malmendier U, Shanthikumar D (2007) Are Small Investors Naive about Incentives? Journal of Financial Economics 85(2):457-489

March JG (1991) Exploration and Exploitation in Organizational Learning. Organization Science 2(1):71-87

Mayer RC, Davis JH, Schoorman FD (1995) An Integrative Model of Organizational Trust. Academy of Management Review 20(3):709-734

McKnight DH, Chervany NL (2001-2002) What Trust Means in E-Commerce Customer Relationships: An Interdisciplinary Conceptual Typology. International Journal of Electronic Commerce 6(2):35-59

McKnight DH, Cummings LL, Chervany NL (1998) Initial Trust Formation in New Organizational Relationships. Academy of Management Review 23(3):473-490

McKnight DH, Choudhury V, Kacmar C (2002). The Impact of Initial Consumer Trust on Intentions to Transact with a Web Site: A Trust Building Model. Journal of Strategic Information Systems 11(3-4):297-323

McNair DM, Lorr M, Droppleman LF (1971) Profile of Mood States. San Diego, CA: Educational and Industrial Testing Service

Milberg SJ, Burke SJ, Smith HJ, Kallman EA (1995) Values, Personal Information, Privacy and Regulatory Approaches. Communications of the ACM 38(12):65-74

Milne GR, Boza ME (1999) Trust and Concern in Consumers' Perceptions of Marketing Information Management Practices. Journal of Interactive Marketing 13(1):5-24

Mitchel RLC, Philipps LH (2007) The Psychological, Neurochemical and Functional Neuroanatomical Mediators of the Effects of Positive and Negative Mood on Executive Functions. Neuropsychologia 45:617–629

Mitchell ML, Mulherin JH (2007) The Impact of Public Information on the Stock Market. Journal of Finance 49(3):923-950

Morales L (2011) Google and Facebook Users Skew Young, Affluent, and Educated. Last retrieved 2013-09-23, http://www.gallup.com/poll/146159/facebook-google-users-skew-young-affluent-educated.aspx.

Nann S, Krauss J, Schoder D (2013) Predictive Analytics on Public Data - The Case of Stock Markets. Proceedings of the 21st European Conference on Information Systems

Neumann R, Strack F (2000) "Mood Contagion": The Automatic Transfer of Mood between Persons. Journal of Personality and Social Psychology 79(2):211-223

Niederhoffer V (1971) The Analysis of World Events and Stock Prices. Journal of Business 44(2):193-219

Nofsinger JR (2005) Social Mood and Financial Economics. Journal of Behavioral Finance 6(3):144-160

Norberg PA, Horne DR, Horne AA (2007) The Privacy Paradox: Personal Information Disclosure Intentions versus Behaviors. Journal of Consumer Affairs 41(1):100-126

Odean T (1998) Volume, Volatility, Price, and Profit When All Traders are above Average. Journal of Finance 53(6):1887-1934

Oh C, Sheng O (2011) Investigating Predictive Power of Stock Micro Blog Sentiment in Forecasting Future Stock Price Directional Movement. Proceedings of the 32nd International Conference on Information Systems

Page SE (2007) Making the Difference: Applying a Logic of Diversity. Academy of Management Perspectives 21(4):6-20

Pavlenko A (2008) Emotion and Emotion-Laden Words in the Bilingual Lexicon. Bilingualism: Language and Cognition 11(2):147-164

Pavlou PA, Gefen D (2004) Building Effective Online Marketplaces with Institution-Based Trust. Information Systems Research 15(1):37-59

Phelps J (2000) Privacy Concerns and Consumer Willingness to Provide Personal Information. Journal of Public Policy & Marketing 19(1):27-41

Poetz MK, Schreier M (2012) The Value of Crowdsourcing: Can Users Really Compete with Professionals in Generating New Product Ideas? Journal of Product Innovation Management 29(2):245-256

Prince M (1993) Women, Men, and Money Styles. Journal of Economic Psychology 14(1):175-182

Prosser WL (1960) Privacy. California Law Review 48(3):383-423

Qiu L, Welch I (2004) Investor Sentiment Measures. National Bureau of Economic Research

Rajagopalan MS, Khanna V, Stott M, Leiter Y, Showalter TN, Dicker A, Lawrence YR (2010) Accuracy of Cancer Information on the Internet: A Comparison of a Wiki with a Professionally Maintained Database. Journal of Clinical Oncology 28(15): 6058

Rao T, Srivastava S (2012) Using Twitter Sentiments and Search Volumes Index to Predict Oil, Gold, Forex and Markets Indices. Working Paper

Reagans R, Zuckerman EW (2001) Networks, Diversity, and Productivity: The Social Capital of Corporate R&D Teams. Organization Science 12(4):502-517

Rochet JC, Tirole J (2003) Platform Competition in Two-Sided Markets. Journal of the European Economic Association 1(4):990-1029

Roll R (1981) A Possible Explanation of the Small Firm Effect. Journal of Finance 36(4):879-888

Rosenbaum P, Rubin D (1983) The Central Role of the Propensity Score in Observational Studies for Causal Effects. Biometrika 70(1):41-55

Rotter JB (1971) Generalized Expectancies for Interpersonal Trust. American Psychologist 26(5):443-452

Rousseau DM, Sitkin SB, Burt RS, Camerer C (1998) Not So Different after All: A Cross-Discipline View of Trust. Academy of Management Review 23(3): 393-404

Rouwenhorst KG (1998) International Momentum Strategies. Journal of Finance 53(1):267-284

Ruiz EJ, Hristidis V, Castillo C, Gionis A, Jaimes A (2012) Correlating Financial Time Series with Micro-Blogging Activity. Working Paper

Rysman M (2004) Competition between Networks: A Study of the Market for Yellow Pages. Review of Economic Studies 71(2):483-512

Saunders EM (1993) Stock Prices and Wall Street Weather. American Economic Review 83(5):1337-1345

Scharfstein DS, Stein JC (1990) Herd Behavior and Investment. American Economic Review 80(3):465-479

Schwartz N, Clore GL (1983) Mood, Misattribution, and Judgments of Well-Being: Informative and Directive Functions of Affective States. Journal of Personality and Social Psychology 45(3):513-523

Schwarz N (1990) Feelings as Information: Informational and Motivational Functions of Affective States. In Sorrentino RM, Higgins ET (eds): Handbook of Motivation and Cognition: Foundations of Social Behavior, Vol. 2, pp. 527-561. Guilford Press, New York

Schwind M, Hinz O, Stockheim T, Bernhardt, M (2008) Standardizing Interactive Pricing for Electronic Business. Electronic Markets 18(2):161-174

Semiocast (2013) Half of Messages on Twitter Are Not in English, Japanese is the Second Most Used Language. Last retrieved 2014-06-22, http://semiocast.com/downloads/Semiocast_Half_of_messages_on_Twitter_are_not_in_English_20100224.pdf

Shapiro C, Varian HR (1998) Information Rules: A Strategic Guide to the Network Economy. Harvard Business School Press, Boston.

Sharpe W (1966) Mutual Fund Performance. Journal of Business 39(1):119-138

Shiller RJ (2002) Bubbles, Human Judgment, and Expert Opinion. Financial Analysts Journal 58(3):18-26

Shiller RJ (2003) From Efficient Markets Theory to Behavioral Finance. Journal of Economic Perspectives 17(1):83-104

Silveira V (2012) Taking Steps to Protect Our Members. Last retrieved 2013-09-23, http://blog.linkedin.com/2012/06/07/taking-steps-to-protect-our-members

Simmons JP, Nelson LD, Galak J, Frederick S (2011) Intuitive Biases in Choice versus Estimation: Implications for the Wisdom of Crowds. Journal of Consumer Research 38(1):1-15

Simonsohn U (2010) eBay's Crowded Evenings: Competition Neglect in Market Entry Decisions. Management Science 56(7):1060–1073

Singh T, Hill ME (2003) Consumer Privacy and the Internet in Europe: A View from Germany. Journal of Consumer Marketing 20(7):634-651

Smith HJ, Milberg SJ, Burke SJ (1996) Information Privacy: Measuring Individuals' Concerns about Organizational Practices. MIS Quarterly 20(2):167-196

Smith HJ, Dinev T, Xu H (2011) Information Privacy Research: An Interdisciplinary Review. MIS Quarterly 35(4):989-1015

Solove DJ (2006) A Taxonomy of Privacy. University of Pennsylvania Law Review 154(3):477-560

Spann M, Skiera B (2003) Internet-Based Virtual Stock Markets for Business Forecasting. Management Science 49(10):1310–1326

Spiekermann S, Grossklags J, Berendt B (2001) E-Privacy in Second Generation E-Commerce: Privacy Preferences versus Actual Behavior. In: Proceedings of the 3rd ACM Conference on Electronic Commerce, New York

Sprenger TO, Tumasjan A, Sandner PG, Welpe IM (2013) Tweets and Trades: The Information Content of Stock Microblogs. European Financial Management

Statman M (1999) Behavioral Finance: Past Battle and Future Engagements. Financial Analysts Journal 55(6):18-27

Straub DW, Collins RW (1990) Key Information Liability Issues Facing Managers: Software Piracy, Proprietary Databases, and Individual Rights to Privacy. MIS Quarterly 14(2):143-156

Suh B, Han I (2003) The Impact of customer Trust and Perception of Security Control on the Acceptance of Electronic Commerce. International Journal of Electronic Commerce 7(3):135-161

Sunden AE, Surette BJ (1998) Gender Differences in the Allocation of Assets in Retirement Savings Plans. American Economic Review 88(2):207-211

Surowiecki J (2004) The Wisdom of Crowds. Doubleday, New York

Sy T, Cote S, Saavedra R (2005) The Contagious Leader: Impact of the Leader's Mood on the Mood of Group Members, Group Affective Tone, and Group Processes. Journal of Applied Psychology 90:295-305

Taft R (1955) The Ability to Judge People. Psychological Bulletin 52(1):1-23

Tetlock PC (2007) Giving Content to Investor Sentiment: The Role of Media in the Stock Market. Journal of Finance 62(3):1139-1168

Thaler RH (1987) The January Effect. Journal of Economic Perspectives 1(1):197-201

Tirunillai S, Tellis GJ (2012) Does Chatter Really Matter? Dynamics of User-Generated Content and Stock Performance. Marketing Science 31(2):198-215

Treynor JL (1987) Market Efficiency and the Bean Jar Experiment. Financial Analysts Journal 43(3):50-53

Trombley MA (1997) Stock Prices and Wall Street Weather: Additional Evidence. Quarterly Journal of Business and Economics 36(3):11-21

Tsai J, Egelman S, Cranor L, Acquisti A (2011) The Effect of Online Privacy Information on Purchasing Behavior: An Experimental Study. Information Systems Research 22(2):254-268

Tseng KC (2006) Behavioral Finance, Bounded Rationality, Neuro-Finance, and Traditional Finance. Investment Management and Financial Innovations 3(4):7-18

Tucker C, Zhang J (2010) Growing Two-Sided Networks by Advertising the User Base: A Field Experiment. Marketing Science 29(5):805-814

Tversky A, Kahneman D (1991) Loss Aversion in Riskless Choice: A Reference-Dependent Model. Quarterly Journal of Economics 106(4):1039-1061

Van Eck M, Nicolson NA, Berkhof, J (1998) Effects of Stressful Daily Events on Mood States: Relationship to Global Perceived Stress. Journal of Personality and Social Psychology 75(6):1572-1585

Vittengl JR, Holt CS (1998) A Time-Series Diary Study of Mood and Social Interaction. Motivation and Emotion 22(3):255-275

Vul E, Pashler H (2008) Measuring the Crowd Within: Probabilistic Representations within Individuals. Psychol Sci 19(7):645-647

Wann D, Dolan T, Mcgeorge K, Allison J (1994) Relationships between Spectator Identification and Spectators' Perceptions of Influence, Spectators' Emotions, and Competition Outcome. Journal of Sport and Exercise Psychology 16(4):347-364

Watson WE, Kumar K, Michaelsen LK (1993) Cultural Diversity's Impact on Interaction Process and Performance: Comparing Homogeneous and Diverse Task Groups. Academy of Management Journal 36(3):590-602

Weinberg P, Gottwald W (1982) Impulsive Consumer Buying as a Result of Emotions. Journal of Business Research 10(1):43-57

Welch I (2000) Herding among Security Analysts. Journal of Financial Economics 58(3):369-396

Westin A (1967) Privacy and Freedom. Atheneum Books, New York

Whaley RE (2000) The Investor Fear Gauge. Journal of Portfolio Management 26(3):12-17

Williams KY, O'Reilly CA (1998) Demography and Diversity in Organizations. In: Staw BM, Sutton RM (eds.) Research in Organizational Behavior (20), JAI Press, Stamford, pp. 77-140

Wolfers J, Zitzewitz E (2004) Prediction Markets. Journal of Economic Perspectives 18(2):107-126

Wong A, Carducci BJ (1991) Sensation Seeking and Financial Risk Taking in Everyday Money Matters. Journal of Business and Psychology 5(4):525-530

Woodman RW, Ganster DC, Adams J, McCuddy MK, Tolchinsky PD, Fromkin H (1982) A survey of Employee Perceptions of Information Privacy in Organizations. Academy of Management Journal 25(3):647-663

Worthington A (2009) An Empirical Note on Weather Effects in the Australian Stock Market. Economic Papers: A Journal of Applied Economics and Policy 28(2):148-154

Wright WF, Bower GH (1992) Mood Effects on Subjective Probability Assessment. Organizational Behavior and Human Decision Processes 52:276-291

Ye S, Wu SF (2010) Measuring Message Propagation and Social Influence on twitter.com. Social Informatics 6430:216-231

Yoon E, Guffey HJ, Kijewski V (1993) The Effects of Information and Company Reputation on Intentions to Buy a Business Service. Journal of Business Research 27(3):215-228

Yuen KS, Lee T (2003) Could Mood State Affect Risk-Taking Decisions? Journal of Affective Disorders 75(1):11-18

Zellner A (1962) An Efficient Method of Estimating Seemingly Unrelated Regressions and Tests for Aggregation Bias. Journal of the American Statistical Association 57(298):348-368

Zhang X, Fuehres H, Gloor P (2010) Predicting Stock Market Indicators through Twitter – "I hope It Is Not as Bad as I fear". In: Collaborative Innovations Networks Conference, Savannah, GA

Zhu F, Zhang X (2010) Impact of Online Consumer Reviews on Sales: The Moderating Role of Product and Consumer Characteristics. Journal of Marketing 74(2):133-148

Zuckerman M (1984) Sensation Seeking: A Comparative Approach to a Human Trait. Behavioral and Brain Sciences 7(3):413-71

Printed in the United States
By Bookmasters